THE STORY BEHIND

Copyright © 2018 Emily Prokop.

Published by Mango Publishing Group, a division of Mango Media Inc.

Cover and Layout Design: Elina Diaz

Mango is an active supporter of authors' rights to free speech and artistic expression in their books. The purpose of copyright is to encourage authors to produce exceptional works that enrich our culture and our open society.

Uploading or distributing photos, scans or any content from this book without prior permission is theft of the author's intellectual property. Please honor the author's work as you would your own. Thank you in advance for respecting our author's rights.

For permission requests, please contact the publisher at:

Mango Publishing Group
2850 Douglas Road, 3rd Floor
Coral Gables, FL 33134 USA
info@mango.bz

For special orders, quantity sales, course adoptions and corporate sales, please email the publisher at sales@mango.bz. For trade and wholesale sales, please contact Ingram Publisher Services at customer.service@ingramcontent.com or +1.800.509.4887.

The Story Behind: The Extraordinary History Behind Ordinary Objects

Library of Congress Cataloging

ISBN: (print) 978-1-63353-828-3 (ebook) 978-1-63353-829-0
Library of Congress Control Number:
BISAC category code: REF023000—REFERENCE / Trivia

Printed in the United States of America

THE STORY BEHIND

The Extraordinary History Behind Ordinary Objects

EMILY PROKOP
Producer, *The Story Behind* Podcast

Coral Gables

To Layla and Cameron:

May you always remain as curious as you are now about the world around you.

CONTENTS

INTRODUCTION

Call it curiosity. Call it distraction. Call it ADHD. But I've always found it difficult not to wonder about the history of the things around me. When the I got the internet in my teenage years, a whole new world opened up to me. Sure, I had the library before, but I wanted answers to my questions and I wanted them now.

You may be looking at this book and wondering if you might feel the same way. Maybe you don't consider yourself a fan of history—I know I never did, aside from the few history teachers in high school who found a way to make it interesting. I was never good at remembering dates or even locations (I should probably list our Google Home as a co-author for answering me all those times I had to check the years of a particular battle or era). But what did stick in my head was trivia—the details that made history fun and exciting to me.

While I'm no expert, I did love talking about the trivia I learned, usually during dinner-party conversations when my socially awkward self couldn't hold back the latest new piece of information I'd learned or some fascinating anecdote I'd come across. Luckily, podcasting existed, so I got to spout off all the random trivia that had been swirling in my brain for years into my show, *The Story Behind.*

Every week, I seemed to stumble across something new and want to share it with people, or I would choose something as simple as a paper clip and, seemingly out of nowhere, find myself in a Google rabbit hole investigating its origins.

Some of the research in *The Story Behind* may seem a bit scattered, but that's because not everything has the very basic story we're familiar with of "white guy invents this, makes millions." I like looking at the thought process the inventors may have had, or even how things were before their inventions.

Many times, there are objects that are impossible to really find the history of, but the theories and educated guesses are just as fun to talk about, and so I included a few.

Millennials/Gen-Xers like me (apparently I'm part of a micro-generation known as Generation Oregon Trail) may feel like we're living in quite possibly the worst timeline, where the economy is uncertain, politicians are lying to us, and it seems radical groups are running amok and gaining notoriety in the process. Sometimes, researching episodes of *The Story Behind* helps me keep in perspective all the advances in technology and how inventions like those mentioned in the book have made our lives better.

So let's dive into *The Story Behind*.

P.S.—I won't be offended if you read it in the bathroom.

PART ONE:
AT THE OFFICE

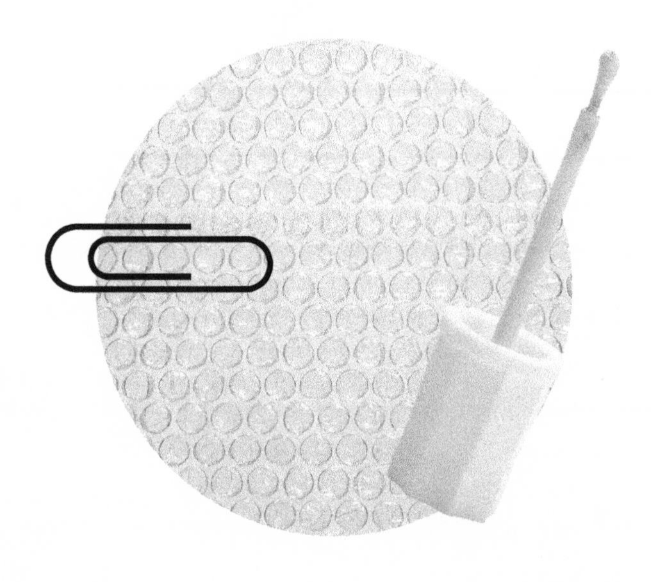

BUBBLE WRAP

When World War II soldiers returned from the war and got an education, thanks to the GI Bill, this helped them purchase houses. The country boomed, not only with new babies, but also with new homes being built. The 1950s saw a shift from more traditional decor to families looking to the future, with space exploration on the horizon and futuristic movies and television influencing style choices.

By that time, wallpaper had evolved from a luxury only the rich could afford to a more affordable commodity anyone could use to easily and quickly spruce up their walls. In 1957, Alfred W. Fielding and Marc Chavannes wanted to create a high-end plastic wallpaper. The main draw would be that the wallpaper was textured and would add a pop (pun totally intended) of fun to the walls.

In Fielding's garage, they sealed two plastic sheets together, creating air pockets trapped inside, and put a paper backing on it. Unfortunately, the design wasn't as popular as they had hoped, and the two inventors had this amazing material but no use for it. Originally, it was known as Air Cap, and Fielding and Chavannes formed the company Sealed Air to market it.

They first tried marketing it to greenhouses as a sort of cheap insulation, but it was difficult to market plastic for walls. Luckily, another futuristic innovation would help them use this material.

IBM premiered its new 1401 variable-word-length computer in 1959, but there was concern about the difficulty of shipping the new hardware without damaging it using the traditional shipping materials of newspaper, straw or horsehair.

Frederick Bowers, a marketer at Sealed Air, pitched the material to IBM, and finally a use was found for Bubble Wrap. (Bubble Wrap, by the way, is the trademarked name.) Sealed Air began expanding its product offerings to more shipping materials, such as envelopes made with Bubble Wrap padding, which became especially popular in the 1980s with the popularity of the floppy disk. (The first ones were actually floppy and easily damaged.)

The days of sealing two plastic sheets together were long gone, but, in 1957, a machine was made to produce the material with the bubbles evenly spaced. The machines used today are not that different, although there are more of them and they are much bigger than the original, which was the size of a moderately-priced sewing machine at the time.

The materials used now are more environmentally friendly, but still remain strong enough to reuse. However, newer shipping material is always being explored by companies like Amazon who ship huge quantities of items, since big rolls and sheets of Bubble Wrap take up a lot of storage space.

One question that comes up a lot with Bubble Wrap is about how to properly use it to package items—with the bubbles out or in. The recommended way is having the bubbles facing inward to better pad the item being wrapped. It also helps keep small parts in place more effectively.

But what many consider Bubble Wrap's best quality is the stress relief that comes from popping it. And that stress relief isn't just a theory. It was shown in a 1992 study that subjects who were given Bubble Wrap to pop were found to be more relaxed and alert afterward. A few theories exist as to why, including one in which our primitive brain associates the sensation with crushing ticks or insects, but the more plausible (and less gross) theory is that humans are drawn to tactile (touch) sensations, and using worry beads or fidget toys, or popping Bubble Wrap, can help us release that stress.

DID YOU KNOW?

Sealed Air licenses a calendar version of Bubble Wrap, in which each day is printed on a piece of paper underneath Bubble Wrap and consumers can pop a bubble a day.

TL;DR VERSION

- Bubble Wrap was originally created as 3-D wallpaper.
- It took a while before the inventors of Bubble Wrap found a use for it, but once computers became more mainstream, manufacturers sought better ways to ship them, and Bubble Wrap proved to be the perfect solution.
- Popping Bubble Wrap has been proven to be stress-relieving, so pop away!

COMIC SANS

Ah, the font many typographers and graphic designers cringe at. If you don't know the font I'm referring to, think of the speech bubbles in comics. The hand-drawn-looking characters became the basis for the Comic Sans font, which has become associated with schools and kids' activities, and, if you collected Beanie Babies in the '90s, you'll recognize it as the font from the TY company.

It's definitely not as formal as Times New Roman, a common font in newspapers, or Helvetica, which is used for many business names and logos. It's fun and light-hearted. But that doesn't come without downsides, as many who are familiar with fonts find it difficult to take anything seriously if it's written in Comic Sans.

Vincent Connare was the brains behind the font. While working at Microsoft in 1994, he was tasked with finding a font for the speech bubbles of a character in a computer program called Microsoft Bob, which was designed for kids.

This software interface was made to look like a cartoon, but the designers were using Times New Roman in the speech bubbles, and that didn't seem right to Connare. He looked to the stack of comic books he had in his office for inspiration, particularly *Watchmen* and *The Dark Knight Returns*. He began crafting the letters, but, when applied to the speech bubbles in the software, the all-capital-letter font wouldn't fit.

He added lowercase letters to the font, but it was too late, and Microsoft Bob had already come and gone by then. But Microsoft did pick up the font for its Movie Maker application. The original name was Comic Book, but it was then changed to Comic Sans.

The font was kept by Microsoft and added to the system fonts in Windows 95 and Internet Explorer. Apple seemed to copy the concept and included the similar font called Chalkboard in Mac OS X.

If you've been around enough people who work in design, you probably know that Comic Sans gets a pretty bad rap, despite being pretty common in everyday signage.

Dave and Holly Combs, graphic designers from Indianapolis, started a group called Ban Comic Sans in 1999 after their employer insisted they use the font for a children's museum exhibit.

When Lebron James left the Cavaliers to "take his talents" to the Miami Heat in 2010, Cavaliers owner Dan Gilbert wrote an angry letter on the team's website in Comic Sans. Though many were angered by James leaving, the intensity of the letter written by Gilbert was dulled by the whimsical (or, if we're being less generous, ridiculous) nature of the font he used.

When the European Organization for Nuclear Research, known as CERN, discovered the Higgs boson, otherwise known as the God particle, in 2012, the discovery was almost overshadowed by the disgust people expressed after the scientists presented their findings in Comic Sans.

Two French designers, Thomas Blanc and Florian Amoneau, started a Tumblr called the Comic Sans Project, in which they reimagined famous brand logos redone in Comic Sans, and it's a little off-putting to see something as iconic as the AC/DC band logo made into a playful-looking logo, just by changing the font.

When Pope Benedict XVI stepped down, the Holy See published a digital scrapbook filled with photos of his papacy, and committed the graphic designers' "deadly sin" of using Comic Sans as the typeface.

Even Weird Al Yankovic poked fun at the font in his song *Tacky,* singing, "Got my new resume/it's printed in Comic Sans."

I've asked teachers why schools use Comic Sans, and the best answer I've gotten (aside from, "it's just how we've always done it") is that the font is the most similar to how letters should be written by their students when they're learning to write. It's one of the few fonts that show the lowercase *a*'s, especially, as children are taught to write them, while other fonts have a hook at the top of the *a*.

Even as I write this in Google Docs, I'm offered 16 fonts for my use, and only two of them don't have that hook on the top of the *a*. Want to guess which ones? Corsiva and Comic Sans.

Comic Sans has also been said to be a recommended font for those with dyslexia, as it might be easier for them to read fonts that look more like handwriting, and the heaviness of it may make it stand out more. But studies on this have varied as far as how true this may be.

No matter what your personal opinion of Comic Sans is, you may want to heed some comic-book advice: With great power comes great responsibility. Use Comic Sans wisely.

DID YOU KNOW?

When you see Sans in a font name, it refers to "Sans Serif" (or "without serifs"). Serifs are the tiny little lines attached to the ends of letters, such as in Times New Roman, whereas a font like Arial doesn't have serifs.

TL;DR VERSION

- Comic sans was created by Vincent Connare when he was working at Microsoft in 1994.

- It was originally designed for Microsoft Bob, but wasn't finished in time for its release. It premiered with the Windows Movie Maker application instead.
- When Lebron James left the Cavaliers to play for Miami in 2010, Cavaliers owner Dan Gilbert wrote an angry letter on the team's website in Comic Sans, which made it lose a bit of its edge.

CORRECTION FLUID

Once upon a time, long, long ago, there was no such thing as autocorrect (gasp!) or even a backspace key (double-gasp!). Back in a time when typewriters, ink, and paper were the norm, people were tasked with typing or hand-writing important papers and correspondence. Oh, sure, there were erasers for mistakes made with pencil, but, for formal documents, erasers sometimes just made more of a mess.

Aside from misspellings, typewriters were known to jam or even type something on the wrong line. Even though typing classes were common in high schools, you can still imagine how easy it was to make mistakes when you think about how long you've been typing on a computer keyboard, yet still have to use the backspace button.

In 1951, when single mom Bette Nesmith Graham found work as an executive secretary for W. W. Overton, the Chairman of the Board of the Texas Bank and Trust, she realized she was prone to these types (pun intended) of mistakes in her typing. And she wasn't the only one. Her colleagues were also making more mistakes when the bank began using electric typewriters.

(Now, kids, remember, this story ends happily, but this is where the scary part comes in.)

Back then, when a mistake was made on a typed piece of paper, the typist would have to take the whole piece of paper out of the typewriter and (eek!) throw it away and start from scratch! All over one error!

One day, Graham came upon painters decorating bank windows for the holidays. When the painters made a mistake, Graham noted, they would just cover it with an additional layer of paint, instead of trying to redo it. This gave her an idea.

She went home and began experimenting with white tempera paint in her kitchen, because of its fast-drying and long-lasting properties. Graham began using bottles of her creation at work to correct her typing mistakes, and as soon as other secretaries saw what was called Mistake Out, by 1956, orders started pouring in.

Graham realized she was onto something, especially with the new popularity of electric typewriters, which made people more prone to mistakes because of their higher finger sensitivity compared to manual typewriters.

With some help from her son's chemistry teacher, Graham experimented with the formula and settled on a combination of titanium dioxide, which is commonly used as a pigment enhancer in paints and even some makeup, and mineral spirits, which allows the product to dry faster.

She renamed her invention Liquid Paper and began learning business management and marketing to help promote her product, while still working as a secretary.

But not for long.

Graham was soon fired from her job for, of all things, a typing mistake. She accidentally typed her own company's name at the top of a letter, instead of the bank's name. But by that time, it didn't matter. Liquid Paper had already taken off, and Graham could afford to go full-time with the business.

She patented the Liquid Paper formula and even pitched the product to IBM in 1957, but was turned down. But, by the next year, Graham's factory was filling orders of ten thousand bottles per day, and businesses realized there was a market for the product.

In 1966, Wite-Out hit the market as a competitor and became the generic name for correction fluid. Graham was finally able to sell the product in 1979 to Gillette a year before her death. While computers have taken the place of most typewriters, the popularity of Liquid Paper for instances when someone makes a mistake with a pen and paper still keeps the product (and its competitor) in business.

DID YOU KNOW?

Bette Nesmith Graham, the inventor of correction fluid, was the mother of Michael Nesmith, a member of the popular '60s television show and band the Monkees. He is known for his signature green hat and came into his audition for the show with a load of laundry to fold while he waited.

TL;DR VERSION

- Prone to typing mistakes, Bette Nesmith Graham invented correction fluid in her kitchen.
- IBM turned Liquid Paper down, not seeing a market for it.
- Wite-Out came later to the market, even though the name eventually became the generic term for correction fluid.

THE PAPER CLIP

Many of the objects I'm writing about in *The Story Behind* evolve in one way or another from the original invention—but not this one. In fact, before paper clips, papers were held together using string or wax. It was common for office clerks to have a number of small cubbyholes at their desks to keep corresponding papers together.

There were a number of methods used to keep papers together, like pasting pages together, melting sealing wax, or using a needle and thread to sew through slits in the papers made with a pen knife.

One of the most popular methods was the use of straight pins made of iron in the nineteenth century, but, as you can imagine, people were easily pricked. The common problem with these methods was the damage done to the pages.

By the mid-1800s, steel became a more common material to work with, since it was easily malleable but would still hold its strength. It also had the characteristic of not rusting, which made it an ideal material to explore for binding papers instead of the iron straight pins, which would leave rust marks on papers.

Many believe the paper clip was invented in Norway by Johan Vaaler. In fact, a giant-paper-clip monument was erected in Oslo to commemorate the invention. However, the paper clip Vaaler invented doesn't resemble the ones we know today (and the monument to him isn't even his original clip design).

Vaaler's design, created in 1899, was more like a rectangular version of the paper clip we know today, but with the inner loop removed.

Paper clips continued to hold a special place in Norway's history, though. Going back to World War II, it was decreed illegal to fasten any sort of patch to one's clothing, especially if it showed allegiance to the king. Norwegians began fastening a single paper clip to their clothing as a way to show solidarity.

Before that, however, another designer, named Samuel B. Fay, had designed a clip that was meant to be used to fasten tickets to articles of clothing. It had the shape of an upside-down teepee, a triangle with the two angled sides extending beyond the point at which they met. But, again, still not quite the paper clip we know today.

The modern-day paper clip isn't actually patented. But the machine to make them is. The official name of the paper clip we're most familiar with is the Gem Paper Clip. No one actually knows who invented the very first one, although there has been incorrect speculation that it came from the Gem Manufacturing Company in Britain in the 1870s.

While the design came over to the United States in the 1890s, it wasn't until 1899 that a patent was issued to William Middlebrook of Connecticut for the machine that would take a thin piece of steel and loop it around twice to form the Gem clip.

This design was springy enough to hold papers together, and the additional loop meant that taking off the clip wouldn't result in tearing the paper as much as Vaaler's invention did. He sold his patent for the design to the office-supply company Cushman & Denison, which trademarked the name Gem Paper Clip.

While variations have come along, the original design has lasted more than one hundred years and shows no sign of going away anytime soon.

The paper clip is pretty ubiquitous by now. Even if we don't see a physical one on a daily basis, we see the symbol for it on the button to attach something to an email.

Aside from holding papers together, a Canadian by the name of Kyle Macdonald found another use for it. He found himself with no job, but he had a website and a red paper clip. And, in the early 2000s, he knew that the craze over buying, selling, and trading on eBay could be his ticket to a house.

So he put up a website with the intention of trading his paper clip, with the goal of trading and trading up to finally get a house. At first, he started by trading his red paper clip for a pen shaped like a fish, then the pen for a door knob, then the door knob for a barbecue. He went through about fourteen trades with other things, like a live appearance with Alice Cooper and a paid spot in a movie, to finally end up trading for a house in Saskatchewan.

But probably the most notable paper clip of all would be Clippy. Or Clippit, which is his formal name.

When Microsoft introduced its Office Assistant Clippy, some might say he was a predecessor to customer-service bots and even our smart home devices. He was actually a kind of reincarnation of Microsoft Bob, which helped new PC users navigate Windows. Microsoft Bob (which we talk about in "The Story Behind Comic Sans"), though, with its simple graphics, was made fun of for being childish and was quickly scrapped in 1995.

Microsoft tried to revive the idea of an assistant with Clippy, who began popping up in Microsoft Office in 1997. Its creator, Kevan Atteberry, was actually contracted by Microsoft to design Clippy, which, funnily enough, he did on a Mac.

For those who weren't around back then, if you opened up a Microsoft Word document and began typing the word "Dear," for example, a little anthropomorphized paper clip would appear in the bottom-right-hand corner with a word bubble that would say something to the effect of, "It looks like you are writing a letter. Would you like some assistance?"

Sure, people could disable Clippy, but the fact that he was on by default angered people. In fact, some of the earliest memes I remember were people using Microsoft Paint to superimpose obnoxious phrases into Clippy's word bubble, like having him say "Sometimes I watch you sleep," since having something like a digital assistant was not as common as it is today.

Eventually, Microsoft realized Clippy was not as popular as they had hoped, and by 2002, he was disabled by default. In 2004, Microsoft began running an anti-marketing campaign to kill him, believe it or not, and then, in 2007, he was finally pulled from Microsoft Office software.

DID YOU KNOW?

Similarly to the way Norwegians fastened paper clips to their clothing during World War II as a symbol of solidarity, in more recent times, the safety pin attached to one's clothing became a symbol of support in the United Kingdom for refugees following the vote to leave the European Union in 2016, as well as in the United States following the election of Donald Trump in the United States for those who wanted to identify themselves as allies for women, immigrants, and LGBT and minority groups.

TL;DR VERSION

- Before paper clips, papers were held together by tying a string around them, using wax or glue, or using a needle and thread to sew pages together.
- Johan Vaaler of Norway is celebrated as the inventor of the paper clip, but it's not the same paper-clip design we know today, even though the statue commemorating his design depicts it so.
- The old Microsoft Office assistant in the shape of a paper clip is known as Clippy, for the most part, but his formal name is Clippit.

THE PINK SLIP

In the majority of my research on this, one name kept coming up: Peter Liebhold, a curator at the Smithsonian Institution's National Museum of American History in Washington, DC., who has been searching for the origin of the pink slip, only to come across several dead ends, unfortunately.

There was one story that placed the origin back in the early days of the Ford Motor Company, when supervisors put white papers in some workers' cubbyholes to show they were doing a good job and pink papers in the cubbyholes of other workers as a sign they were on their way out the door.

However, the originator of that tale was a management consultant who had heard the tale in college and couldn't actually verify its accuracy, but passed it to numerous others.

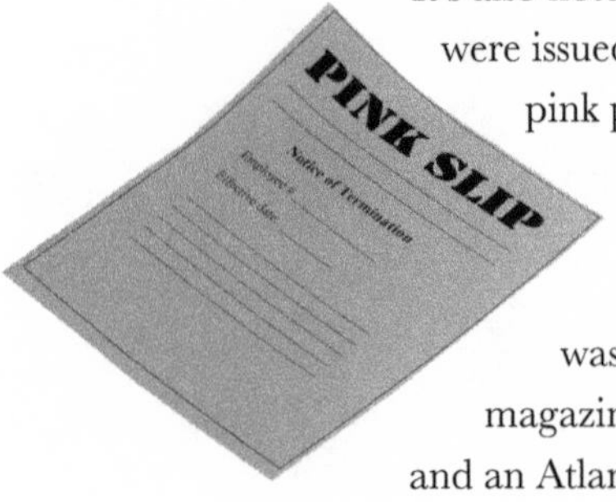

It's also noted that, in 1904, warnings were issued to typographers on pink pieces of paper, and if a typographer received too many, he would be fired. Another reference to them was a rejection letter from a magazine printed on pink paper, and an Atlanta newspaper published a line in 1906 that says, "There is nothing like a prospective pink slip to fill the brawny athlete with zest and ginger."

The Oxford English Dictionary referenced one of the first instances of the phrase being used in a pulp novel called *Covering The Look In Corner*, by Gilbert Patten, about baseball in 1915, when it was mentioned by a character saying, "And have Murphy hand me the pink slip tonight."

So, while we might never find the original company known for using pink slips, the phrase remains in our lexicon. But it's not just slang, as it's actually used on termination documents in the workplace today.

Back in 2008 and 2009, when the US experienced a rise in unemployment, it became fashionable to hold Pink Slip Parties on the day of a mass layoff. Discounted pink champagne and pink cocktails were served, and some parties even evolved to organizers inviting recruiters. Luckily, since 2010, unemployment has steadily decreased and these parties are not as common.

Even though we may not know about the first pink slip in existence, the phrase is generally known throughout the US, while, in Germany, being fired is referred to as getting the blue letter, and, in the French military, the personnel office is given a yellow paper.

Some surmise that the pink slip has nothing to do with the color of the paper, but that the word "pink" means to cut ties, or literally pierce or stab, as in the term "pinking shears." Pink also has the definition of piercing by means of ridicule, and one of the anecdotes claimed the paper was pink so that other employees would plainly see who was getting the axe.

But we do know the origin of the term "pink slip" when it comes to car titles, as it was used in the movie *Grease*.

"Pink slip" referred to the color of the car title document in California until 1981 (or 1988, depending on the source). The car title was a small, four-by-five-inch piece of paper until California switched to a larger rainbow title, which was harder to forge. It has since changed back to pink, but now has a blue border.

DID YOU KNOW?

While Liebhold still hasn't been able to track down the origin of the term "pink slip," he did find that secretaries used to bundle documents in red twine, giving rise to the term "red tape."

TL;DR VERSION

- The origins of the term "pink slip" are hard to pinpoint, although one story points to Ford Motor Company putting either white or pink papers in the cubbyholes of workers, with white signifying they had done a good job, while pink meant they had done poorly.

- Following the mass layoffs of 2008 and 2009, Pink Slip Parties were thrown by terminated employees.
- "Pink slip" also refers to car titles since, up until the 1980s, the paper they were printed on was pink.

PART TWO:
FASHION

BUTTONS

Before they were known as the go-to fastener, historically, buttons were more decorative. Early buttons have been found dating back to 2000 BC, and some say they've been around even longer, though it's the early Romans who are said to have been among the first to use the button as a clothing fastener by looping material over it. The Romans are also known to have created another fastener called the fibula, which is a precursor to today's safety pin.

Buttons have been made out of almost every material imaginable. Buttons came to the western world with British soldiers returning from the Crusades, bringing with them intricately painted buttons made by the Turks and Mongols. Medieval fashionistas were inspired by these, and the gowns of the upper class began to be adorned with their own versions.

It wasn't until a few years later, when fashion began to favor a more tailored silhouette, that the button became more of a fastener. It allowed dressmakers to create looks that could cling tightly to the body, accentuating the waist and even tightening sleeves so they wouldn't have to be sewn together in the morning and cut open at night.

Soon, button materials were assigned to different social classes, with the most prosperous being able to have the more opulent materials such as precious metals or gems, while rosary makers were allowed horn, bone, or ivory. Wood or cloth buttons were relegated to the lower class. The various materials shifted among classes as time went on, although those in power or the wealthy were still able to get the fanciest buttons, and were more likely to wear them for decoration, in addition to using them as fasteners.

Collectors look at the eighteenth century as the Golden Age of buttons, since they were more intricate and larger then. Many depicted carvings of ancient myths or characters in folklore. Buttons also went from more of a ball shape to becoming flatter, making them perfect canvases for painters of the era.

When Queen Victoria mourned her late husband Prince Albert's death in 1861, she started the trend of wearing buttons made out of jet, many with her VR monogram, which stood for Victoria Regina. These became known as Mourning Buttons. Her subjects took notice, and demand for them grew.

Another Victorian button trend was the use of lithographs and portraits of high-profile figures and celebrities. International marvels such as the Eiffel Tower were stamped onto metal buttons as well.

Buttons were used in almost every facet of fashion, including on shoes, requiring a tool called a button hook to help people fasten them. You may remember a scene in the 1939 movie *The Little Princess* in which Shirley Temple's character, Sara, has trouble using her button hook.

But the most common buttons were the ones with four holes. This left little room for decoration, but these buttons were more practical and stayed put on clothing longer than those with attachment loops on the backs.

With the arrival of the zipper in 1913, button sales started to decline; however, they proved useful during World War II, as many buttons used on uniforms were made such that, if a pilot was shot down, he could remove a button and use it as a compass, either because it was in fact a tiny compass, or because it was made to magnetically point north when suspended by a string. The insignia and varied stampings on military buttons also make them popular among button collectors today.

DID YOU KNOW?

Joseph Coors Jr.—of the Coors beer family—decided to create an indestructible button in 1989. He ordered one made by the ceramics department of the family's brewing company. Zirconium oxide was used to make Coors' button harder and much stronger than steel.

TL;DR VERSION

- Buttons have been around for thousands of years, although the majority of that time they were mostly for decorative purposes.
- During the Victorian era, button makers began making flatter buttons and painting or carving tiny works of art on them. These are still among the most coveted by collectors.
- Before zippers and other closures, buttons were used on so many articles of clothing that a special device known as a button hook was used for shoes and gloves.

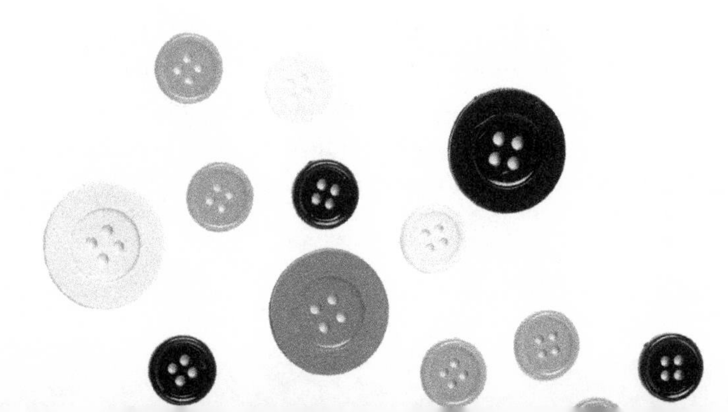

MUSTACHES

Mustaches have gone in and out of fashion throughout history. The styles have evolved over time, with one of the more popular types coming from the art of the 1600s, specifically portraits done by Sir Anthony Van Dyck of King Charles I, who sported a pointed goatee and handlebar mustache. Sometimes this combination is referred to as a Charlie, but, more often than not, it's called a Van Dyck.

Toward the end of the century, facial hair went through another dip in popularity—so much so that, when Tsar Peter of Russia returned from his tour of Western Europe in 1698 he is said to have pulled out a razor and begun shaving guests at his welcome-home party, and decreed faces must be kept clean-shaven to emulate the hairless faces in the West. After a while, a beard tax was imposed on those who wished to maintain their facial hair.

Mustaches were back in style during the Victorian era and were so popular that a new utensil was invented: a spoon with a mustache-shaped cutout over the bowl to protect one's mustache from food.

They fell out of favor again by the Depression, when advice guides warned

job seekers against showing up to work with mustaches, which was said to give a bad, unkempt impression. By World War II, not only did clean-shaven soldiers set the trend, but the once-popular toothbrush mustache (yeah, Adolf Hitler's look) fell out of favor and never made a comeback, as you can imagine, despite America's love of Charlie Chaplin.

It seems weird that it was this long ago, but William Howard Taft was the last American president to wear a mustache, though Presidential nominee Thomas Dewey came close to depriving Taft of that honor when he ran against Franklin Roosevelt in 1944 and Harry Truman in 1948.

Mustache trends come and go, but many classics are still commonly seen, like Teddy Roosevelt's walrus mustache (which is also a favorite of actor Wilford Brimley) or the iconic chevron mustaches known to grace the faces of Freddie Mercury, Burt Reynolds, and Tom Selleck, or even Hulk Hogan's signature horseshoe mustache (not to be confused with the Fu Manchu, which reaches down to the chin in a similar fashion, but is only connected to the face above the lip).

By the way, the Fu Manchu was named after Dr. Fu Manchu, a character most often portrayed as an evil genius in novels, movies, radio, and comics in the first half of the twentieth century.

The mustache's fall from favor around World War II included the world of baseball. Satchel Paige was the last major-league baseball player to sport a mustache during a game, in the 1940s, until Reggie Jackson came to his 1972 spring training with the Oakland A's with what teammates called a "scraggly" mustache.

Rollie Fingers, then a teammate of Jackson's, conferred with his fellow players, and it was decided that they would all grow mustaches, hoping the facial hair would get banned by management, which would force Jackson to shave as well—but it backfired. Owner Charlie Finley and manager Dick Williams offered players $300 each to grow mustaches by opening day.

It's said Rollie became so attached to what became a waxed handlebar mustache that, when he was offered a position with the Cincinnati Reds in 1986, he instead chose to retire because the Reds refused to exempt him from their no-facial-hair policy.

In 2003, two friends in Melbourne, Australia, were talking fashion trends in a bar and decided it was time the mustache became fashionable again. They decided to create a fundraising campaign and emailed 30 of their friends with a proposal for each of them to chip in $10 toward prostate cancer awareness and prevention and to grow a mustache for the month of November. With that, Movember was born.

Over the next year, the initiative grew, and Movember became official. Since then, every November, more men raise money and awareness—now not only for prostate cancer, but for testicular cancer and men's mental health as well. More than $730 million Australian (about $562 million US) has been raised through the Movember efforts.

They project that, by 2030, they'll reduce premature deaths of men by 20 percent.

DID YOU KNOW?

Alex Trebek said, "What is Goodbye?" to his classic mustache back in 2002; however, whenever Will Ferrell portrays him on *Saturday Night Live*, he is always shown with a mustache.

TL;DR VERSION

- The mustache continually goes in and out of fashion, and a beard tax was enacted in Russia during the 1600s when Tsar Peter decided to shave his party guests' faces to emulate the western world.
- During the Depression, clean-shaven faces were recommended to job seekers.
- Movember was started by two friends in Australia after they talked about bringing the mustache back and hoping to raise money and awareness for prostate cancer.

NAIL POLISH

In 3200 BC, Babylonian soldiers were known to stain their nails before battle as part of their war paint and strategy for intimidating their enemies. But the earliest known paint for nails was developed in China around 3000 BC, using a mixture of beeswax, egg white, gelatin, gum arabic, and vegetable dyes. This was used to color-code people by their rank in society. As the centuries went on, nail decoration became a way to signify that the wearer didn't have to engage in manual labor, thus was more well-off than others.

Gold and silver were used by the highest classes, while black and red were used by warriors, and pale colors by the lower classes.

By the fourteenth century BC, creating designs with henna on hands became a popular ritual. Queen Nefertiti was known for popularizing its use on hands and feet, but it was Cleopatra, in the first century BC, who was said to popularize painting only her nails. She preferred dark red, while common folk were only allowed to sport more muted or pale colors.

As far as historians know, not much changed over the years with nail polish, although, when you look at classic paintings throughout the Middle Ages and Renaissance, you'd be hard-pressed to find one of a female with painted nails. By contrast, in the 1400s, Aztecs and Incas used nail art as totems for battle, using sticks and natural dyes to draw eagles on their nails.

Even though actual nail polish may not have been used, nails and hands were very well-taken-care-of, especially for those in the upper class. King Louis' manicurist is said to have developed the first nail file from a dental tool.

Salons also became popular in Paris in the late 1800s, using creams, oils, and powders to clean and shine the nails. In fact, the term "polish" came from the act of shining the nails during this time. In 1878, finding herself divorced and in need of money, Mary Cobb opened the first manicure parlor in America, in Manhattan.

Believe it or not, it was the invention of nitrocellulose, created as an explosive in the 1830s, that became the basis for nail polish today. Cellulose was mixed with nitric acid, originally to make a smokeless gunpowder, but it could also be used to make lacquers which would harden.

The brand Cutex used nitrocellulose to manufacture nail polish, although it wasn't consistently pigmented until later.

Combining nitrocellulose with colored dye, carmakers found this material perfect for painting cars. But when a French makeup artist named Michelle Ménard, who worked for an automobile company, saw this, she thought it would make a great paint for fingernails. She brought her idea to the Charles Revson company, and, together, they began manufacturing nail polish in the 1920s. The company soon changed its name to Revlon, which we know it as today.

Even through the Depression, nail polish was priced affordably enough to make women feel like they could indulge in a little bit of a luxury. Thanks to the trend set earlier by Cobb's manicure parlor, women could treat themselves in their homes to a freshly painted set of nails.

By the end of World War I, fashion skewed toward bolder choices for women, throwing aside the more conservative looks of past decades. This included nail polish. Aside from the more vivid colors of nail polish being offered, another trend in nail painting was the "moon manicure" or "half-moon manicure," in which nails were grown out and only half were painted, leaving the top unpainted in a crescent shape. In the early twentieth century, nails were kept clean, short, and rounded, but, by the 1930s, the moon manicure was seen on nails that ended in a sharp point. Actresses Dorothy Flood and Joan Crawford were known for sporting this style.

Acrylic nails were actually developed in 1957 by a dentist named Fred Slack Jr., who cut his thumbnail accidentally. In a pinch, he used a combination of aluminum foil and dental acrylic to fix the nail and realized it would be the perfect solution for women who couldn't grow the long nails that were in fashion. With his brother Tom, he developed the first acrylic nail extensions.

There are two men who claim to have invented the French manicure. For those who don't know, a French manicure is a natural-looking pink or beige color on the nails with a painted white tip. The first to claim the idea is none other than Max Factor, who created it for women of Paris in the 1930s who wanted a clean look. For those who have had French manicures, you know the white tips make hands look perpetually like they are freshly washed.

But the more well-known story behind the French manicure was that a Hollywood makeup artist named Jeff Pink was stuck with the problem of constantly having to paint and repaint actresses' nails for each day of shooting scenes, back in 1975. Since movies are not shot in chronological order, nail polish had to be changed accordingly. He came up with the French manicure not only for actresses on set, but also for runway models in Paris, for a look that would go with their constantly changing outfits.

Nowadays, nail polish can be found in almost any shade imaginable. New trends emerge with nail polish and nail art every year, with celebrities leading the way. One of the most popular shades of Chanel nail polish is called Rouge Noir, a mixture of dark red and black thought to look like dried blood, which was popularized by Uma Thurman in *Pulp Fiction*.

DID YOU KNOW?

There have been a few instances over the years of people "piercing" their nails and wearing tiny charms or jewelry in the holes. The trend was started by American actress Titania, in 1897, who wore diamond-studded rings through her long nails, but while some predicted it would become the next big thing, it never really took off.

TL;DR VERSION

- Painting hands and nails has been around since 3000 BC.
- While nail maintenance has been a constant trend, it wasn't until the twentieth century that innovations in automobile painting led the way to nail polish in different shades.
- The origin of the French manicure is contested between Max Factor in the 1930s and a makeup artist named Jeff Pink in 1975.

POCKETS

Do you ever wonder why women's pockets are usually much smaller than men's? That was actually how pockets began. Men had pouches that acted as pockets sewn into their breeches, vests, or overcoats as far back as the seventeenth century, but, if a woman wanted a place to keep her belongings safe while traveling outside the house, she had to have special pouches tied with ribbon underneath all those layers of skirts and petticoats. To get to those, she would need to lift up all the layers and (scandalously) show some skin.

Since it was rare for a woman to be the money-keeper back then, what did she keep in her pockets? It seems these weren't necessarily the small pockets we would recognize today. Women were said to have carried pin cushions and sewing supplies; snuff boxes; smelling-salts bottles, since fainting was a common occurrence; and even cakes and bonbons.

It became common for slits to be cut in skirts for women to be able to access these pouches easily, but, thanks to Marie Antoinette popularizing slimmer, Grecian-cut dresses, the big-skirts-and-dresses fashion declined, taking pockets with it, in favor of a slimmer silhouette. That's when outside purses were introduced—and the "fad" hasn't gone away.

As clothes tailored with pockets increased, you can imagine the number of pickpockets did as well. Though in general a nuisance, pickpockets hold a little bit of charm in popular culture, from the Artful Dodger in Charles Dickens' *Oliver Twist*—picking pockets in a time when to be caught doing so could be a one-way ticket to the gallows—to an illusion master today, billed as theatrical pickpocket Apollo Robbins, who famously stole everything except the guns from Jimmy Carter's Secret Service detail in 2001. That included the president's top-secret itinerary and all the keys to the motorcade.

Harry Houdini even wrote a book in 1906, called *The Right Way to Do Wrong*, with a passage about various ways pickpockets were able to accomplish their crimes: "A common pickpocket trick is for the operator to carry a shawl or overcoat carelessly over the left arm, and to take a seat on the right side of the person they intend to rob in a streetcar or other vehicle.

"Sometimes a small and very sharp knife is used to cut the side of the dress or pantaloons of the victim, so that the purse may be abstracted without going into the pocket directly. Others of this light-fingered gentry wear light overcoats with large pockets removed. They will endeavor to stand near a person, preferably a woman, who is paying her fare and has displayed a well-filled purse. The pickpocket then carelessly throws his coat over her dress, and by inserting his hand through the outside opening of his own pocket, quietly proceeds to abstract her purse."

Now, if you're thinking you probably won't have to worry about anyone slashing your pantaloons, more modern ways to pickpocket rely mostly on diverting attention, such as using an accomplice to distract the victim, known as the "mark," while the pickpocket has a seemingly valid excuse to get close to them, like the sandwich technique, for example.

This is when the accomplice walks in front of the mark while the pickpocket walks behind them both. When the accomplice stops short, inevitably the mark crashes into them with the pickpocket then crashing into the mark, providing an excuse to get close enough to grab the contents of the mark's backpack, purse, or pockets with a quick sleight of hand.

Luckily, pickpocketing has been on the decline for the last half-century, at least in America. Mandatory schooling has kept potential thieving children off the streets, but busy tourist destinations are still some of the most popular spots for pickpockets.

One of the latest trends in pockets, though, has come about in dresses with the braggable detail of having them sewn in.

You may have seen internet memes making fun of this, but as a woman, I can tell you we really do get excited when our dresses come with pockets. (Just ask those who complimented me on my wedding dress—the first thing I said was "Thank you! It has pockets!")

If you're a male, you may not realize just how little pocket space women have. Sometimes, we may buy a pair of slacks only to find out the pockets are fake or sewn closed. Again, this goes back to that silhouette Marie Antoinette popularized. Since pockets are usually placed on the hips, the fashion industry (and women trying to emulate the models on the runway) shy away from adding any more bulk to that area.

Yet, with the technological advances that led to bigger mobile phones, the demand for pockets has increased. And, as a personal plea, would it be so much to ask for cargo pants to make a comeback?

Then again, purse makers have a lot to gain from the fashion industry taking its time adding more functional pockets to women's clothing. I joked with my coworker about this as we were talking about her dress and we decided to rally against the purse industry, or Big Prada as we're now calling it.

DID YOU KNOW?

Major William P. Yarborough played a very big role in Army airborne operations during World War II. Dissatisfied with the standard paratrooper jumpsuit, he not only helped design better jump boots, but also introduced cargo pockets to the uniform to hold additional supplies.

TL;DR VERSION

- While it was common for men to wear pouches as pockets sewn to their slacks in the seventeenth century, women's pouches were usually a separate item that was fastened underneath their layers of skirts.
- Pickpocketing has steadily declined as mandatory schooling became more popular in the United States.
- Women's pockets are generally smaller or fake because the fashion industry tends to shy away from adding any extra bulk to the hips.

TATTOOS

Otzi the Iceman is the oldest mummified human found, dating back to 5,300 years ago. Not only did he show evidence of pierced ears, but he also had at least fifty-seven tattoos.

The tattoos consists of dots and small crosses from his ankles to his neck. These are thought to be medical tattoos, as opposed to decorative, since they correspond to joints that would have been strained throughout his lifetime.

Thousands of years later, the tradition of tattoos continued with Egyptians, especially females. The mummified women were found to have tattoos on the tops of their thighs, breasts, and stomachs, and this is thought to have been done for protection during pregnancy and labor.

Tattoos have been popular for centuries all over the world, with evidence found in Greece, Italy, China, Japan, and the Polynesian islands, where the word "tattoo" comes from.

It comes from the Tahitian word *tatau*, meaning to hit or strike. The practice was brought to Europe following Captain James Cook's expedition in 1769. When missionaries came to the islands and converted natives to Christianity, they tried to stop the practice of tattooing. It was thought to damage the body that was supposed to be kept a temple, but banning it was unsuccessful. Tattoos even spread in popularity, especially as a good-luck charm among workers with high-risk jobs, like miners or sailors.

It's pretty much a toss-up who is more superstitious, actors or sailors. However, aside from the occasional real knife replacing a stage knife, acting doesn't come with as many fatalities as sailing does. Being out on the open sea for months or years at a time can really test one's patience and faith. It's no wonder sailors—and also military personnel serving at sea—sought anything that gave them a bit more luck or to remind them of how far they'd traveled, like a tattoo of a hula girl for Hawaii or a dragon for China. Tattoos of swallows were earned every five thousand nautical miles traveled, and anchors indicated the sailor had crossed the Atlantic Ocean.

Tattoos for luck included foot tattoos of pigs and roosters, which were shipped in crates during World War II; in the case of a shipwreck, these crates would float, and the animals would survive. Tattooing them on a sailor's feet was done in hope that the sailor would also survive a shipwreck.

You've most likely seen nautical stars more often, as they're a particularly popular tattoo. That and a compass rose were symbols for a sailor of finding the way home. Next time you take a trip to Macy's, you can tell your friends the red star in the logo is part of the founder, R. H. Macy's, tattoo he got as a young boy on a whaling ship.

The man most known for iconic tattoo designs, especially those associated with the sea, is Norman Collins, better known as Sailor Jerry.

Influenced by Eastern art and technique, he made a name for himself in Hawaii, developed his own pigments and influenced generations of future tattoo artists, including Ed Hardy. He was also responsible for popularizing single-use needles and using an autoclave to sterilize the instruments.

Before the tattoo machines we have today, Native Americans and people of the Polynesian and Hawaiian islands used sharpened bones or rocks to carve designs into flesh, and then filled the wounds with soot from burnt candlenut shells or natural dye.

The intricate and large tattoos associated with Samoan culture were made using a bamboo cutting tool with a comb that was dipped in ink and pierced the skin with the designs. A similar procedure was used in the nineteenth century, with gypsies in Egypt tying several needles together and pricking the skin, then filling it with the soot from burnt wood or oil mixed with breast milk.

When Thomas Edison invented the electric pen in 1876, it used a small motor to move a needle up and down on a piece of paper as a person was writing, creating a stencil for copying.

It didn't catch on, unfortunately, but when tattoo artist Samuel F. O'Reilly saw the electric pen, he realized this could be a much faster way to tattoo the skin, since the standard tattoo artist could only puncture the skin two or three times per second, while the machine quickened the process up to fifty times per second. In the 120 years tattoo machines have been used, there has been very little in the way of differences between O'Reilly's original idea and what tattoo artists use today.

DID YOU KNOW?

Actor Sean Penn is a tattoo artist, and one of his clients was comedian David Spade, who sports a Calvin tattoo (from the *Calvin and Hobbes* comic strip).

TL;DR VERSION

- Tattoos were first used in medicine on strained joints and also on women for fertility.
- Nautical tattoos remain popular and each symbol represents a sailor's journey or feat.
- When Thomas Edison's electric pen didn't take off, Samuel F. O'Reilly saw its potential as a tattoo machine.

THE T-SHIRT

Maybe you remember seeing comical underwear in old cartoons, called a union suit. This was usually a red flannel one-piece with legs and arms and—probably what it's most known for—a butt flap. Basically, it's the winterized bro-romper of yesterday.

In the late 1800s, workers would cut their union suits at the waist and would only wear the top half when it got too warm, which is considered the beginning of T-shirts.

By the turn of the century, clothes makers had found a way to create a neck hole that didn't require buttons and could fit over one's head. The key to this innovation was the ability for the neck hole to snap back without getting stretched out.

In 1905, the US Navy made the plain white cotton undershirt part of its required uniform, and those working in the engine rooms of ships could fashion the shirts for comfort. By World War I, the undershirt had made its way into the official uniform of the US Army. Around that time, the word "T-shirt" first appeared in print in F. Scott Fitzgerald's *This Side of Paradise.*

There is a rumor that the T-shirt fell out of favor after the 1934 movie *It Happened One Night* featured a scene in which Clark Gable unbuttons his shirt to reveal a naked chest, but there's no factual evidence to support the decline. In fact, it was another movie from the same decade that debuted what's considered the first appearance of a graphic T-shirt. In the movie, three men are seen stuffing straw into a scarecrow while wearing green T-shirts with the word "OZ" written on them. That movie, of course, is *The Wizard of Oz.*

High-school jocks were responsible for bringing T-shirts to the younger generation, since they were worn to keep football padding from chafing. But it was Marlon Brando's portrayal of Stanley Kowalski in *A Streetcar Named Desire* in 1951, that became most associated with wearing a T-shirt as outerwear. Soon after, the T-shirt became a blank fabric canvas for businesses and brands to showcase their logos.

It's not just businesses that like to emblazon their logos on T-shirts. There are plenty of T-shirt trends throughout the years. Like the Che Guevara shirt, for example.

Marxist revolutionary Che Guevara's face became a popular design for T-shirts because of his association with overthrowing an oppressive government in Cuba. But he was also part of a movement to herald in Fidel Castro's communist government instead. Those who wear his face on T-shirts as a form of rebellion and individualism may not know that it was Che who would round up rebellious youth on the streets and put them to work in camps.

Like many trends that find their way into the zeitgeist, wearing the Che Guevara shirt became *passé*, especially when Urban Outfitters pulled merchandise with his face in 2012 after an open letter condemned them for promoting a figure who represented anti-democratic views and the repression of millions under the communist regimes in Cuba and around the world.

Another notable T-shirt came about when fashion designer Katharine Hamnett designed the white T-shirt with big black letters that says CHOOSE LIFE seen in Wham!'s music video for *Wake Me Up Before You Go-Go*. Hamnett said she wanted her design to be copied, and similar large-font slogans on shirts continued to be popular, notably as part of a short-lived 2011 trend with an episode of *Glee* called "Born This Way," in which the characters used the shirts to highlight their unique attributes.

Back in middle school, I remember begging my parents for a $40 T-shirt from Calvin Klein with the little C and big K it seemed everyone had. When I saved up enough birthday money, I went out and bought it, only to have the cool kids in school tell me I was a poser for wearing it, since I didn't even own Calvin Klein jeans.

If $40 sounds like a lot for a T-shirt—which it is, don't get me wrong—that's nothing compared to the public mockery that ensued when Kanye West teamed up with French fashion brand APC for a clothing line featuring a plain white tee called the Hip-Hop T-Shirt, priced at $120. If that's too steep for you, the following year, he debuted the Plain T-Shirt, which, again, was a plain white T-shirt retailing for "only" $90.

When you watch a major sporting event like the Super Bowl or World Series, you may notice the winning team celebrating their victory and donning championship hats or T-shirts with their team's logo right away. The next day, victory garments are available at many major retailers for fans to purchase. It's pretty amazing those just happen to be handy, right? But what happens to the merchandise made for the losing team's non-existent victory that was also on hand?

It used to be burned and destroyed, since no one would want to buy an inaccurate shirt that said "Atlanta Falcons 2017 Super Bowl Champions"—that is, unless you really, really hated the Patriots and only acknowledged the first half of the game—but, today, somewhere in the world, people are wearing hats and shirts proclaiming the Falcons the winner because of a group called World Vision, which collects unwanted sports attire and distributes it to people in disaster areas and impoverished nations.

When the Colts lost the Super Bowl in 2010, the team's inaccurate victory merchandise was sent to Haiti, which had just been hit by an earthquake.

Only about 10 percent of clothing donated in America gets put back on American shelves at second-hand retailers like Goodwill or the Salvation Army. A huge proportion of it is compressed into bales and shipped to developing or impoverished countries to be bought at a low price, then put out for sale in street markets, including all those free promotional T-shirts you may have gotten.

While this is seen as good for those street sellers, the low price of American second-hand clothing has put factories and hand-made retailers in those counties out of business. If you're a husky-sized American donating your clothes, they don't do a lot of good in those countries where obesity isn't such a national problem. Those clothes either aren't accepted or they're sold to those with the means and skills to modify the garments to be put up for sale.

Wearing an ironic T-shirt has become associated with hipsters, especially featuring bands "you've probably never heard of," or kitschy slogans and graphics for events they've probably never attended. But Haiti-based photographer Paolo Woods shot a series of photographs showing out-of-place second-hand T-shirts being worn by Haitians, many of which made a full revolution of being made in sweatshops in Haiti, sent to America to be purchased, donated, then returned to Haiti to be sold on the streets.

The photographs included a young boy wearing a shirt that says "This is my Lucky Shirt," which is stained and full of holes, and a young girl with dark hair in a shirt that says, "Kiss me, I'm a blonde."

Then again, there are people like my husband, who holds on to every T-shirt he gets (until they "mysteriously" disappear when I do his laundry).

DID YOU KNOW?

The word "T-shirt" comes from its shape, and was added to Webster's Dictionary in 1920.

TL;DR VERSION

- Workers in the late nineteenth century were known for cutting their union suits when it was too hot and wearing the top portion only, which is considered the first known T-shirt.
- T-shirts became part of the standard navy and army uniforms, and then became popular with kids and young adults when high school footballers began wearing them to keep their pads from chafing.
- Merchandise is made for both teams in the Super Bowl, but the losing team's is usually donated to an impoverished or disaster-stricken area.

PART THREE:
FOOD

BUBBLE GUM

For those of you who have ever gotten distracted at work, consider this story a little form of validation. Walter Diemer began his foray into the bubble-gum business when he worked for the Fleer gum company in Philadelphia. The founder, Frank Fleer, had been trying to come up with the perfect gum for blowing bubbles since the early 1900s, but he had trouble selling it because it was just too sticky—if you can imagine bubble gum being even more sticky than it is today.

But when Diemer was working as an accountant for Fleer, he began experimenting with new gum recipes in his spare time. When he was just twenty-three years old, he found a recipe that was significantly less sticky than previous versions. It also stretched much more easily, making it perfect for blowing bubbles.

He brought a five-pound sample to a local grocery store for others to try, and it ended up selling out the same day. Fleer began marketing Diemer's creation and Diemer himself was in charge of teaching salesmen to blow bubbles to help increase sales. If you've ever wondered why bubble gum was pink, that was the only food coloring available in the factory for Diemer to use, so it became the traditional color of bubble gum.

That bubble gum, which Fleer president Gilbert Mustin named Dubble Bubble, was the first commercially-sold bubble gum. For years, it was the only bubble gum on the market, until another company came onto the bubble-gum scene. But you might not be as familiar with the gum from this company as you are with the baseball cards associated with them now. That company was Topps Chewing Gum, and the bubble gum it produced was none other than Bazooka.

Bazooka came about when four brothers started Topps in 1938 in Brooklyn. They began producing Bazooka gum (named after the instrument, not the weapon) and used a comic strip as its wrapper. The character in the comic, Bazooka Joe, used to be the top of the heap when it came to well-known marketing characters. He was a little kid with an eye-patch, which he was said to wear, not for medical reasons, but to be "more interesting." The comics featured groan-worthy jokes, riddles and puns—basically, the perfect fodder for dad jokes nowadays.

Some examples:

> What does Santa Claus say when he works in his garden? Hoe Hoe Hoe!
>
> What do little men who live under bridge eat? Troll house cookies!

> Why isn't your nose twelve inches long? Because then it would be a foot!

In 1952, Topps began replacing the comic with another gift for chewers: a card featuring Major League Baseball players with their picture and stats listed. When Topps saw the popularity of these cards and how kids would trade them with their friends, they began focusing exclusively on baseball cards and, through the years, expanded into other sports.

Bazooka gum became its own entity, separate from the baseball cards, but a stick was still included in the packs, usually without its own wrapper, which would leave marks on whatever card it was touching. If you remember getting that gum, which is no longer included in packs of sports cards, you may remember that the taste and texture left much to be desired.

The gum went back to featuring Bazooka Joe and his motley crew of friends, revamping their styles as the decades went on, but 2012 saw the end of Bazooka Joe comics when Bazooka rebranded. Blue raspberry was added as an additional flavor the red, white and blue original package was ditched for more saturated hues. Wrappers now feature word games and puzzles and codes to use on the company's websites.

Bazooka Joe had gone from one of the most recognized product mascots to only 7 percent of kids knowing who he was by the 2000s. In case you hold out hope of seeing him again, when former Disney CEO Michael Eisner bought the rights to the character in 2007, rumors began circulating in Hollywood that a movie would soon follow, but we have yet to see it.

Because of what some call the war on sugar, it's no surprise that gum sales have decreased in the past few decades, especially with studies showing a correlation between gum-chewing and an increase in cavities. But gum has also been proven to improve test scores and act as a mental warm-up before taking a test.

A study at St. Lawrence University tested participants who chewed gum before and during testing, and participants who didn't chew anything, and they found those who chewed beforehand were able to recall information faster. The researchers think gum chewing warms up the brain, so to speak, by pumping more blood to the head.

In another study, gum-chewing was said to quicken reaction time and accuracy when participants were tasked with detecting patterns of numbers within a set. The act of chewing gum can be seen as a way to help focus the brain.

But if you're a fidgeter who loves your gum, you might want to find a new vice if you're planning on going to Singapore. Gum has been banned there for more than twenty years, unless you have a prescription.

One more thing: Let's put to bed the claim that gum stays in your stomach for seven years if swallowed. This probably stemmed from packaging labeling gum as indigestible. This doesn't mean it doesn't go through your system—it only means your body can't break it down the way it does other foods, and it comes out pretty much the same as it goes in.

DID YOU KNOW?

Bubble gum was included in soldiers' rations in the 1940s.

TL;DR VERSION

- An accountant at the Fleer gum company used his free time to try to come up with a recipe for gum that could be used for blowing bubbles.
- Topps sold bubble gum before it began selling baseball cards.
- Chewing gum can help focus the brain for tests or memorization.

CHOCOLATE CHIP COOKIES

If you've ever looked at an online list of things that were invented accidentally, no doubt you would find chocolate chip cookies pretty high on the list.

The story you most likely heard was that the owner of the Toll House Inn was making cookies and ran out of nuts for her signature nut cookies. Or that she was planning on making chocolate cookies and ran out of baker's chocolate and used pieces from a Nestle chocolate bar instead, which didn't mix with the dough the same way baker's chocolate did. You may have even heard that chocolate pieces accidentally fell into a batch of cookie dough and the rest is history.

It's such a fun story, and everyone seems to love a happy accident. It's no wonder it's usually found on those clickbait-y listicles. (Sometimes, it's even the one "you'll never believe," as the headline assures you.)

But it wasn't.

When you find out more about Ruth Wakefield, you realize this wasn't a woman who was prone to in-a-pinch substitutions or accidentally dropping something in her cooking, only to shrug and say, "Oh well, I guess I'll bake it anyway."

Wakefield studied household arts in the 1920s and began work as a dietitian and food lecturer. To start crumbling that cookie of a common myth, someone so well-versed in culinary arts would probably already know that a chocolate bar wouldn't react the same way as common baker's chocolate when mixed with dough and baked.

When she and her husband, Kenneth, purchased the two-hundred-year-old Toll House Inn in Whitman, Massachusetts, in 1930, they kept up the inn's tradition of serving colonial dishes. It became known for not just its food, though. Ruth Wakefield was known for training her staff to be meticulous about how customers were served.
It was said new waitresses were trained for three months before being assigned the maximum two tables to attend to at the restaurant.

For anyone who's worked in food service, I'm sure you're familiar with a basic code of conduct that, while sometimes strict, could probably fill one single-spaced piece of paper. Those training to work in the Toll House restaurant, however, were given a seven-page manual, which included not only a pristine dress code, but also specific instructions on memorizing orders and setting the table with silverware placed exactly one thumbprint away from the edge of the table.

Another crumble of the cookie: Wakefield was clearly not a woman who would just drop some chocolate morsels into her cookie batter, shrug and say, "Well, whatever, I'll bake them anyway." She probably also wouldn't run out of something as common to her recipes as nuts or baker's chocolate.

So, on that fateful day in the 1930s, when she was preparing cookies that were actually a side to an ice-cream dessert, she was most likely fully aware of what she was doing, although the extent of the cookie's popularity for decades to come probably wasn't on the forefront of her mind.

In the few interviews she gave, Wakefield talked about how she wanted to change things up from her usual butterscotch nut cookies she had been serving with ice cream. And it was clear she purposely added the chocolate chips to her cookies.

It became an instant success. She published her recipe in various newspapers, and it was even featured on the *Betty Crocker Cooking School of the Air* radio program. Following the publication of Wakefield's book *Toll House Tried and True Recipes* in 1936, these original chocolate-chip cookies proved to be such a scrumptious success that Ruth had no choice but to repeat the recipe. She called her new invention the

"Chocolate Crunch Cookie" and published the recipe in several Boston and New England newspapers. When Ruth's Chocolate Crunch Cookie recipe was featured on an episode of the *Betty Crocker Cooking School of the Air* radio program, the popularity of the humble chocolate chip cookie exploded and the cookie soon became a favorite all across America. The popularity of the cookie further increased after Ruth published the still popular, *Toll House Tried And True Recipes*, featuring the *Toll House Chocolate Crunch Cookie*, in 1936.

Wakefield even approached Nestle about collaborating and gave her recipe to them for free, to be reprinted on packages of chocolate chips and Nestle chocolate bars. It's said Wakefield was given as much Nestle chocolate as she wanted in exchange—not a bad form of payment, if you ask me.

Now, in a time when there seem to be new and more decadent desserts popping up at least once a month, with many of us feeling our teeth ache just seeing the ads turn up on our social media, chocolate chip cookies may just be so ubiquitous at this point that it would be hard to imagine such a simple treat taking off as it did.

Even with rationing during World War II making simple ingredients like butter, sugar, and chocolate difficult to come by, women were encouraged by ad campaigns to support "that soldier boy of yours" by sending a batch overseas to him.

As with many foods that were popular among World War II veterans when they returned home, chocolate chip cookies took their place among other World War II favorites and became a classic part of culinary Americana.

Fast-forward to the present, and the Toll House is no more. After it burned to the ground in 1984, it was not rebuilt and the only trace that remains is a historical marker at a Wendy's fast-food burger joint.

DID YOU KNOW?

Wally Amos was the first black talent agent at William Morris, and he discovered and signed Simon and Garfunkel. After leaving William Morris, he began baking cookies, using his Aunt Della's chocolate chip cookie recipe, and selling them, soon turning the popular cookies into a business on Sunset Boulevard in Los Angeles in 1975, called "Famous Amos."

TL;DR VERSION

- Despite a popular myth, chocolate chip cookies were not invented accidentally.
- The inventor, Ruth Wakefield, was incredibly meticulous about how the Toll House was run,

including having waitresses train for three months before being given the maximum of two tables to serve.

- The Toll House Inn, where Toll House cookies were invented, burned down in 1984. As of this writing, the spot is now occupied by a Wendy's.

DR PEPPER

This chapter is based on an episode of The Story Behind *podcast that was part of a series on pop-culture references in the movie* Forrest Gump.

Charles Alderton was a young pharmacist in 1885, working at Morrison's Old Corner Drug Store in Waco, Texas. When he wasn't mixing medicine, he was experimenting with the flavored syrups for the soda fountain. He kept a journal of the various combinations he came up with, and when he finally found one he liked, he offered it to the owner, Wade Morrison, and then to customers, who loved the drink and began ordering it by its nickname, the Waco. It was Morrison who named the drink Dr Pepper, supposedly after an old colleague, but the Dr Pepper museum has collected more than a dozen stories for the name and none can be verified.

The soft drink took off, and Alderton and Morrison began selling the syrup to other pharmacies in the area. Soon, supply couldn't keep up with the demand. When Alderton said he wanted to stay a pharmacist, he suggested Morrison and beverage chemist Robert S. Lazenby develop the drink further.

In 1904, the drink made its World's Fair premiere. If you could travel back in history, this would be probably be the World's Fair to attend—it was the same World's Fair that featured hot dogs and hamburgers served on buns for the first time, as well as ice cream cones.

Research began to surface in the '20s and '30s that showed sugar increased energy and that people's energy levels were subject to crash at certain points during the day. Using this information, Dr Pepper held a contest for an ad slogan, with the winner being, "Drink a bite to eat at 10, 2 and 4."

Its popularity grew even more when it was featured on *Dick Clark's American Bandstand*. By the 1950s, the company decided to drop the period after Doctor in the logo. Because of the font, it was often mistaken for "Dr:" Pepper, instead of the common abbreviation for Doctor.

The slogans have changed since then; all have been considered successful. But with all the ads and campaigns the company has had, none really compares to the man who made advertising more of a storytelling art form.

As I mentioned, we may never know if there was a real Dr Pepper. A theory floating around is that the inventor or drug store owner who sold it named it after the father of a girl he was pursuing. But no one has been able to confirm this rumor.

However, it was most likely started by John W. Davis in the 1930s, after he quit the oil business in Texas and moved to Roanoke, Virginia, to open a Dr Pepper plant. His tactics of selling Dr Pepper to the citizens of Roanoke involved about five thousand different promotions, and his most successful was telling six people every day this version of Dr Pepper's origin.

In his story, Morrison was originally from Christiansburg, Virginia, and was in love with the daughter of Dr. Charles T. Pepper. But, unfortunately, even after he named the soda after her father, the girl broke Morrison's heart.

Only three senior employees at Dr Pepper know the twenty-three ingredients that go into the beverage. But in 2009, a manuscript collector in Oklahoma paid $200 for a notebook from a Texas antique store. Hoping to clean it up and sell it on eBay, he opened it up to reveal a sales ledger. But there was also a page with a recipe on it, titled "D Peppers Pepsin Bitter," which he assumed to be the original recipe.

A spokesperson from Dr Pepper has said the recipe, which is only partly legible, bears little resemblance to the actual recipe used for the soft drink, but the recipe found in the book may actually be a digestive, which is used to help medicine taste better, and is made up of flavored syrups and mandrake root.

Dr Pepper was able to spread to more markets when the FDA and a US District Court declared it a non-cola in the 1960s. While most restaurants can only be loyal to either Coke or Pepsi, Dr Pepper could be sold at either. A cola, by the way, is defined by its origins from the kola nut, which Dr Pepper does not have.

Coca-Cola clearly wasn't happy about yet another competitor and came out with Mr. PiBB in 1972. Just to show it wasn't trying to compete with Dr Pepper, a spokesperson for Coke even said he had never tasted Dr Pepper, so he wouldn't even know if there was a similarity. But the introduction of Mr. PiBB actually helped sales of Dr Pepper, since the similar taste of Mr. PiBB just made consumers crave Dr Pepper even more.

The Coke and Pepsi War through the '80s was brutal to any soft drink not associated with either brand. Dr Pepper remained an independent soda producer until Cadbury/Schweppes purchased the company along with 7UP in 1995, adding to its array of beverages like IBC Root Beer and Welch's soft drinks. As Cadbury/Schweppes grew larger, acquiring Snapple and RC Cola in 2000, it changed its branding to reflect the number of beverages available and is now known as the Dr Pepper Snapple Group.

DID YOU KNOW?

Roanoke, Virginia, is known as the Dr Pepper Capital of the world.

TL;DR VERSION

- The often-told story of the inventor of Dr Pepper naming it after the father of a girl he loved is most likely a myth, although the actual origin story of the name remains a mystery.
- There is no period in Dr Pepper because the makers realized the font used on the logo made it look like it was called Dr: Pepper.
- Dr Pepper is considered a non-cola because it doesn't include kola nut in its ingredients.

THE LOLLIPOP

Maybe you think putting candy on a stick was a newer innovation. But the idea for it came about thousands of years ago when ancient Chinese, Arabs, and Egyptians would stick fruits and nuts in honey and attach them to a piece of wood for easier eating. The honey acted like a preservative, but as time went on, honey, and later sugar, was harder to come by.

It wasn't until the seventeenth century in England, when sugar became more accessible, that street vendors in London began selling meat or soft candy on a stick. It was back then that the term "lolly pop" started to gain traction. Lolly, by the way, means tongue and the pop means "slap."

During the Civil War era, candy was put on the tops of pencils—remember, this was before combining the pencil and eraser was as common as it is today.

This is where the story starts taking a few different turns, since tons of candy makers have wanted to take credit for one of the most popular treats.

One origin story comes out of Connecticut. George Smith, co-owner of Bradley Smith Candy Company in New Haven, came up with the idea of putting candy on a stick. (Well, he didn't come up with it as he was more "inspired" by a nearby chocolate caramel candy company that put its treat on sticks.) He went to his partner, Andrew Bradley, with the idea and named it after his favorite racehorse, called Lolly Pop.

There's a similar story that the McAviney Candy Company, again out of New Haven, Connecticut, would use sticks to stir their boiled hard candy, and at the end of the day, either the owner or employees would bring these sticks home to their children, with the candy hardened onto them, and in 1908, these became marketed as "used candy sticks."

Regardless of which story you believe, George Smith did take a patent out on lollipops in 1931, and even on the word "Lolly Pop." But the patent was difficult to maintain, since companies had been using the term frequently, and the spelling varied between ending with a Y and ending with an I.

Another hassle with getting the patent was that Lollipop with an "I" wasn't an original word. The Oxford English Dictionary places the first usage in print in January of 1784 in the London Chronicle newspaper. Additionally, Charles Dickens, known for using common London street slang, used the term "lollypop" in his novels in the nineteenth century. However, the meaning wasn't referring to a candy on a stick, but to something sweet that was popped into one's mouth, again referring to the tongue as a lolly.

By the way, another company out of Racine, Wisconsin, created what was known as the first automated machine that could put hard candy on a stick in 1908. However, they didn't refer to this as a Lollipop, but they did believe they could produce enough to supply the country with lollipops—or whatever they called them—for an entire year.

Another story comes out of San Francisco in 1912, when a Russian immigrant named Samuel Born invented a machine that could also stick hard candy on a stick, and it was aptly called the Born Sucker Machine.

Unfortunately, when the Great Depression hit, lollipops weren't in high demand. Candy companies like Bradley Smith were forced to close down, negating the patent even further. But when a child star named Shirley Temple graced the screen in 1934's *Bright Eyes* and sang "On the Good Ship Lollipop," sales began to rise again.

Lollipops continued to grow in popularity, making their way into more pop culture, like the Lollipop Guild in 1939's *The Wizard of Oz,* and, in 1958, the doo-wop group the Chordettes successfully covered the song "Lollipop," which beat out the original version by Ronald & Ruby on the charts to reach No. 2.

When Spaniard Enric Bernat came up with the idea of a bonbon on a stick in the late '50s, he originally named it

"GOL," as in a soccer ball going into the mouth as a net. The product was renamed Chupa Chupa—*chupa* being Spanish for "to suck"—but he had some trouble with the branding, which he discussed over coffee with one of his friends, who happened to be the artist Salvador Dalí.

It was Dalí who sketched what would become the Lollipop's logo. Dalí even suggested it be placed on top so everyone would see the name. The logo has been slightly altered in more recent years, but it's still pretty similar to Dalí's original 1969 design.

The lollipop was also a popular treat in the show *Kojak*. At the end of solving a crime, the titular character would unwrap a Tootsie Pop and stick it in his mouth.

Speaking of Tootsie Pops, if you've ever heard the rumor that finding a Tootsie Pop wrapper with an Indian and shooting stars on it and mailing it to the Tootsie Roll company will win you a prize of even more Tootsie Pops, I'm sorry to disappoint you by saying it's false. This rumor has been around since 1931, and Tootsie Roll still receives wrappers from people hoping for the prize.

But, according to their website, if you *do* count the number of licks it takes to get to the center of a Tootsie Pop (without biting it), you can let them know on their Facebook page to receive a Clean Stick award from the famed Mr. Owl of the classic commercials.

DID YOU KNOW?

Samuel Born, who created the first machine that could stick candy onto a stick, is also credited with creating chocolate sprinkles.

TL;DR VERSION

- Putting something sticky and sweet on a stick has been around since ancient times, with fruits, nuts, and honey being popular choices in ancient civilizations.
- Even though a patent was taken out on the Lollipop in 1931, it was difficult to maintain due its widespread popularity.
- The Chupa Chupa logo was designed by Salvador Dalí.

PEANUT BUTTER

Despite its name, peanuts are not actually a nut—they're a legume. That's why you may come across someone who has a tree-nut allergy, which includes almonds, walnuts, and cashews, but can still eat peanuts.

Peanut butter can trace its roots back to the eighteenth century and maybe even further. It's said some of the first peanut butter may have come from Aztecs, approximately a half-century ago, who used to mash peanuts to create a paste.

But it rose to prominence in 1895, when Dr. John Harvey Kellogg—the creator of Kellogg's Corn Flakes—patented the process of making peanut butter and marketed it as a protein substitute for those who had trouble chewing.

He gave it to his patients at Battle Creek Sanitarium, located in Battle Creek, Michigan.

Many credit Dr. George Washington Carver as the inventor of peanut butter in the South during the 1910s and '20s. But, although he was said to invent more than three hundred uses for peanuts, he was not the first to produce the peanut spread.

During World War II, meat and butter were scarce, and rationing was put in place to make sure there would be enough food for everyone. This was when peanut butter really became popular.

It made a cheap and easily accessible substitute protein source, and was even supplied to the US military overseas as part of their rations. To make the peanut butter more palatable, GIs added the jelly that was also included in their rations to it, and when the troops came home, so did the love for peanut butter and jelly sandwiches.

Add that popularity to the Baby Boom that followed World War II, and it became a staple in many quick and easy lunches for kids.

Before Jif hit the market in 1958, peanut oil was used to stabilize peanut butter, but Jif switched the standard formula to use more vegetable oils, such as soy or canola. It surpassed Skippy as the market leader because it had more of a molasses taste, supposedly because honey was also added.

But Jif hit a snag when Consumer Reports reported that half a cup of cooking fat was mixed with Jif's peanuts, giving it the creamy quality it was known for. The FDA proposed that peanut butter consist of 95 percent peanuts, and Jif began changing its label to say peanut "spread" instead of peanut butter.

Following this was what the FDA and lawyers refer to as the Twelve-Year Peanut Butter Case, all over the percentage of peanuts required for the product to truly be called "peanut butter."

This case involved twenty weeks of hearings to negotiate the percentage of peanuts for a product to be called "peanut butter," with peanut butter manufacturers asking for 87 percent, while the FDA argued for 90 percent.

A feisty woman, known for her consumer activism, named Ruth Desmond formed the Federation of Homemakers, who took a particular interest in the Peanut Butter Case. She argued that peanut butter should be kept as simple as possible, and made sure her views were known when she showed up at the start of the hearings in 1965.

Lee Avera, a Skippy official, fought back, arguing that trying to quash the advancements made by the food manufacturers would be detrimental, since America had so much to gain.

Following the hearings, in which the FDA's 90-percent peanuts won out, it took another five years for the US Appeals Court to affirm that standard and, twelve years after the initial proposal, it was put into law that anything labeled as peanut butter must be at least 90 percent peanuts.

If you were to go across the pond to the UK, you would find out that peanut butter isn't quite as loved there as it is in the states.

A teacher at Leiths School of Food and Wine surmises that it may be the texture that turns Britons against it, as they find the mouth-coating of it repulsive.

In the UK, mixing their beloved jam with peanut butter is frowned upon, as well.

But America disagrees. The *Huffington Post* asked its readers in September 2014 what makes the best peanut butter and jelly sandwiches, reporting results for the favorite type of jelly or jam, with strawberry jam as the winner; the favorite type of bread, which turned out to be white; and the favorite type of peanut butter, which turned out to be smooth, as opposed to crunchy. A sandwich combination not quite as beloved is the one favored by my father-in-law: peanut butter and mayonnaise. (Yuck!)

Former president Bill Clinton, as well as Elvis, reportedly favored peanut butter and banana sandwiches. And, speaking of presidents, not one but two presidents of the United States have been peanut farmers: Thomas Jefferson and Jimmy Carter.

DID YOU KNOW?

By the time a child in America is eighteen, he or she will have consumed an average of fifteen hundred peanut butter and jelly sandwiches, although that number may be decreasing because of peanut-free school cafeterias.

TL;DR VERSION

- Many falsely attribute the invention of peanut butter to Dr. George Washington Carver. However, it has been around at least since the Aztecs ground peanuts to make a paste.
- Peanut butter was used as a protein source for those who couldn't chew very well, but gained most of its popularity when it was included in military rations and found to be a wonderful accompaniment to jelly sandwiches during World War II.
- There was a twelve-year battle between peanut butter manufacturers and the FDA over the percentage of peanuts required to officially fall under the category of "peanut butter," as opposed to "peanut spread."

PRETZELS

There isn't one concrete story about the origin of the pretzel, but there are a few twisted tales (pun absolutely intended).

The most common story, however, is that an Italian monk wanted to keep the attention of his catechism students back in 610 AD. He rolled out dough and crossed the ends to mimic the position of the students' arms as they should have been crossed during prayer.

The monk named his creation *Pretiola*, which is Latin for "little reward," but they alternatively got the name *Bracellae*, which is Latin for "little arms." You may also recognize the root of the word *Brach* or *Brace* in other words pertaining to arms, such as "bracelet" and "embrace." By the way, the similar Greek root *Brach* or *Brachio* can be found in the word "Brachiosaur," which literally means "arm lizard."

After pretzels were invented, their popularity spread throughout Europe during the Middle Ages. Because of the cheap ingredients, they were commonly given to the poor as nourishment. They were especially popular during Lent, when Christians were forbidden to eat certain foods. So pretzels, with their simple ingredients of flour and water, became associated with Lent and Easter. Pretzels were even baked and hidden for children to find, along with the hard-boiled eggs that are more familiar to us now.

Even though pretzels were invented by an Italian, they're frequently associated with Germany as a favorite food and, in fact, there is a distinction between a traditional pretzel and a German pretzel. Again, there are different versions of the story about how German pretzels were invented, but one of the more common ones is that they were made accidentally.

In 1839, a baker for the Munich Royal Cafe was preparing to bake pretzels, but instead of brushing them with sugar-water, he accidentally used a sodium hydroxide solution that was used as a cleaner for the bakery equipment. Instead of throwing away the batch (because maybe putting cleaning supplies on food isn't the greatest idea), he baked them anyway. What he found when he pulled them out of the oven was a crispy brown crust and a salty taste that he and guests found delicious.

One more pretzel innovation that came from an accident was the invention of hard pretzels, which came about in 1600 when a baking apprentice in Pennsylvania fell asleep and overcooked the batch, which, of course, made his boss angry. In his rage, while yelling at his employee, he took an angry bite of one of the crunchy pretzels. But once he tasted them, he found he loved them. Good thing for the apprentice!

In addition to the origin story about the religious symbolism behind the knotted dough, the holes in the pretzel also became a symbol for Christianity, the three holes representing the Holy Trinity of the Father, Son, and Holy Spirit. Even in non-religious lore, the pretzel came to represent good luck, prosperity, and a long life.

In Germany, children are known to wear pretzel necklaces on New Year's Day. If you've ever wondered where the phrase "tying the knot" came from, it might have something to do with the Swiss tradition of newlyweds breaking a lucky pretzel in the same manner as a wishbone.

Austria is another country in which pretzel lore is found. At Christmas, pretzels were part of tree decorations in the sixteenth century, but in addition to that, legend has it that monks were baking pretzels in the basement of their monastery in 1510, when they heard Ottoman Turks tunneling underground. The monks were able to alert the city so they could thwart the attack and defeat the Turks. The Viennese King awarded the bakers with their own coat of arms, and many signs outside bakeries still depict a lion holding a shield in the shape of the pretzel.

One more piece of pretzel lore is that pretzels came to America on the Mayflower and were used to trade with Native Americans. Even today, pretzels are still one of the most popular snack foods on the market. They have even been notably featured in the sitcom *The Office;* in Season 3, Episode 5, "Initiation," the office grump, Stanley Hudson, finally shows he can be happy about something: Free Pretzel Day.

Also, admittedly, I've never been a *Seinfeld* fan. However, even I know the scene in the episode "The Alternate Side" in which Kramer is given the line, "These pretzels are making me thirsty!" in a Woody Allen movie, which his friends offer advice about how to deliver.

But, humor aside, when George W. Bush sported a bruised cheek in 2002, it was revealed he had choked on a pretzel and fainted, only to wake up to his dogs looking at him with concern. But he was a good sport about it and even joked about it in a speech given the next day.

DID YOU KNOW?

This isn't the first time food was supposedly invented to help occupy young people. A common story about the origin of candy canes is that they were made to resemble the staffs of the shepherds at the birth of Christ, and the candy was made to keep children quiet during Christmas Eve church services.

TL;DR VERSION

- Pretzels are associated with Christianity, even in their origin as a way to keep kids attentive during catechism.

- Pretzels became associated with Easter in Europe during the Middle Ages.
- The phrase "tying the knot" comes from the Swiss tradition of newlyweds breaking a lucky pretzel.

SALT WATER TAFFY

Take a trip to Atlantic City and leave plenty of room in your suitcase, because one of the biggest-selling souvenirs people will expect you to bring back is salt water taffy.

Now, if you're a *Friends* lover, as I am, you probably already know from the show's trivia buff, Ross, that salt water taffy isn't actually made with salt water, but the salt is added after.

The origin story varies about the original salt water taffy, but the most popular one is that in the 1800s, David Bradley's candy stand on the shore of Atlantic City was flooded by waves from a storm. As he was cleaning up the debris, a little girl walked in and asked to buy some taffy. Bradley replied, sarcastically, that he would be glad to sell her some salt water taffy.

Here's where the details of the story get a bit murky: either the girl's mother, or Bradley's mother, or Bradley's sister overhead the conversation and thought it was a clever marketing name, and, from then on, taffy sold at Bradley's became known as salt water taffy.

Bradley never trademarked the name, and soon the popularity increased. This didn't go unnoticed by Joseph F. Fralinger, who was trying to find his next business venture. Fralinger was a bit of a jack-of-all-trades. He had been a glassblower and a bricklayer before going into the candy business.

He had tried selling various things along the boardwalk with little success, like fish, lemonade, and cigars. But when he was given the opportunity to run a taffy shop in 1884, he had the idea to take the local popularity of salt water taffy and sell it to tourists.

He used decorative oyster boxes to pack in a pound of finger-sized taffy that visitors could take home as souvenirs. His original flavors were molasses, chocolate, and vanilla.

Fralinger even admitted he wasn't the first to use the term "salt water taffy" and credited Bradley's candy stand, although Bradley's taffy was also referred to as Ocean Wave Taffy and Sea Foam Taffy.

Soon after Fralinger opened up more stores in Atlantic City to sell the popular treat, a candy company worker named Enoch James came with his sons to the Jersey Shore, claiming he had been making salt water taffy for years.

James' recipe was less sticky and easier to get out of wrappers. While Fralinger's taffy is usually sold in long pieces, James' was known to be round, which made it easier to fit a whole piece in one's mouth.

James also revolutionized the taffy-making process by cooking it in copper kettles and pulling it over and over onto itself using a hook, which made the product lighter. He also invented many taffy- and candy-making machines that are still in use today.

By the 1920s, "salt water taffy" was still an unregistered trademark. A man named John R. Edmiston tried to put both James and Fralinger out of business by filing a successful trademark on the term, but, when James contested it, the US Patent Office struck down Edmiston's ownership.

DID YOU KNOW?

Although not technically salt water taffy, Laffy Taffy gets its name from the jokes on the wrappers, which are sent in by children.

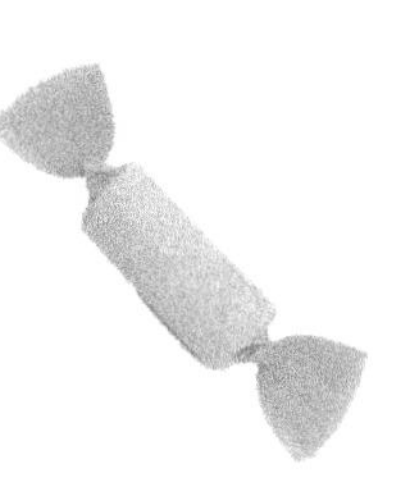

TL;DR VERSION

- Salt water taffy was supposedly a name given to taffy by a candy store owner after a storm flooded his store.
- Salt water taffy became a common souvenir for tourists to bring back from Atlantic City after a taffy salesman started selling one-pound decorative boxes of it.
- Salt water taffy isn't made with actual salt water, although the ingredients do include both salt and water.

SLICED BREAD

If you watched the 1982 movie *Annie,* starring Aileen Quinn, Albert Finney, and Carol Burnett, as many times as I did growing up, you might remember the scene in which Annie is telling Daddy Warbucks how his secretary, Grace, thinks he's the greatest thing since sliced bread. Now, here's a bit of trivia about the movie: When Daddy Warbucks writes the check to the Mudges, he dates it 1933, but the movie they go to see, *Camille*, wasn't released until 1936. Regardless of when the movie takes place, we know that Annie's phrase "greatest thing since sliced bread" doesn't really cover much, since sliced bread was invented only a few years prior, in 1928.

If we're going to get technical, bread has been around since ancient times when wheat grew along the Fertile Crescent and it was discovered that wheat mixed with water and left on its own would rise. And when that dough was baked, it became bread, which some scholars consider one of the factors that turned nomads into settlers, since water was available and bread was an easy way to feed a number of people without using many resources.

It's even said bread was given as payment to workers who built the pyramids in Egypt. Back then, it could be assumed people sliced bread by themselves.

But let's start with the man who invented the bread slicer.

Otto Rohwedder grew up in Iowa, moved to Missouri to become a successful jeweler, then had the idea for pre-sliced bread. He was so taken with his idea, he sold his jewelry stores and moved back to Iowa to begin working on his prototype.

Unfortunately, after a few prototypes were built, the factory he was using burned down, including his machines and the blueprints. But he didn't let that stop him. He kept trying, which ended up working in his favor when the toaster made its debut in 1926. Yes, the toaster came before sliced bread, believe it or not.

And, unsurprisingly, it wasn't very popular until Rohwedder perfected his machine and pre-sliced bread was finally available for the public.

As you may know, when you have sliced bread and leave it out, the open end will go stale quite quickly. Rohwedder needed to find a way to remedy this, so he looked for ways to keep the bread together. Unsliced bread was usually wrapped in cotton and kept in a bread box, but cotton wouldn't help when it came to keeping the slices fresh.

Rohwedder's first attempts failed, including using hat pins or u-shaped pins to keep the slices together, making the bread look more intact.

When he began using waxed paper and cellophane to wrap the sliced bread, keeping the loaves intact, sliced bread finally became viable, with pre-sliced bread outselling unsliced bread within a short period of time. Plus, there were no more pins stuck in slices of bread.

Rohwedder's new bread-slicing machine included a mechanism to wrap the loaves, and, after that, it was hard for him to keep up with the demand from bakeries for his new machine. One of his earliest models is on display at the Smithsonian Institution in Washington, DC.

During World War II, the US was beginning to put rationing in place, and the steel used in making bread slicers became scarce, prompting a ration put on pre-sliced bread. Apparently, this was as irritating to Americans as gas rationing.

Time magazine commented on it in an article that included the passage: "US housewives…vainly searched for grandmother's serrated bread knife, roused sleepy husbands out of bed, held dawn conferences over bakery handouts which read like a golf lesson: 'Keep your head down. Keep your eye on the loaf. And don't bear down.'

Then came grief, cussing, lopsided slices which even the toaster refused, often a mad dash to the corner bakery for rolls."

Luckily, after two months, the ration on sliced bread was lifted. Maybe it was then that the phrase "greatest thing since sliced bread" gained traction, since it was a luxury that was no longer taken for granted.

By the way, if you're wondering what the greatest thing was before sliced bread, just know that while sliced bread was invented in 1928, Betty White was born in 1922.

DID YOU KNOW?

Wonder Bread was the first brand of sliced bread to be sold nationwide.

TL;DR VERSION

- Bread has been around since nomads settled along the Fertile Crescent, and it can be assumed that slices were made by hand.
- Otto Rohwedder invented the machine that slices bread and it gained nationwide popularity.
- Sliced bread became scarce during World War II, but the public outcry prompted the government to lift the rationing after only a few weeks.

PART FOUR:
FUN & GAMES

CROSSWORD PUZZLES

As a child in Liverpool during the Victorian era, Arthur Wynne played a word game called Magic Square, in which a list of words was given and the solver had to arrange the words within a pattern of squares to be read both vertically and horizontally.

The violinist-turned-journalist had come to America and was working at the *New York World* in December of 1913 when his editor asked if he could create a word game for that Sunday's entertainment section. Remembering Magic Square, he created a diamond-shaped grid, but instead of listing words to use, he added the extra challenge of giving clues. He called his game Word-Cross.

Soon, he began making the grids in different shapes and patterns, adding black squares to separate words. It's a said a typographical error was behind the switch from Word-Cross to Cross-Word, then eventually the hyphen was removed. The fad took off, and, within a decade, more and more newspapers picked up the

puzzle for their readers. Librarians complained that avid crossword fans were hogging their dictionaries.

A compilation of past *New York World* puzzles was put together, and the first crossword puzzle book was published in 1924 by Simon & Schuster.

However, not everyone was a fan. The *New York Times* published a scathing column, bashing the new craze, calling it "a sinful waste"—but six years later, they changed their tune and published their first crossword puzzle in 1930.

Crossword puzzles became a mainstay in many newspapers, especially during World War II, when publishers wanted to insert a bit of fun alongside the otherwise tragic news stories. But crossword puzzles also played another role in World War II.

Famously depicted in the 2014 Benedict Cumberbatch movie *The Imitation Game*, a competition in 1914 to solve a *Daily Telegraph* crossword puzzle in under twelve minutes was held in London. Those who were able to beat the clock were sent a summons a few weeks after the competition requesting they report to Bletchley Park. The real reason for the competition was to find minds adept at solving intricate puzzles, which the military thought could be useful in cracking Nazi codes. Those who served in this capacity have since said that the skills needed to solve crossword puzzles don't lend themselves that much to understanding the German language.

By the 1950s, the crossword puzzle became a daily part of newspapers. Meanwhile, across the pond, where Wynne had originally played his word games as a child, another form of crossword became equally popular. It was based more on wordplay and ambiguity in its clues. Lyricist Stephen Sondheim published a version of this in *New York Magazine* in 1968.

The first American Crossword Puzzle Tournament was held in 1978, and it still continues today with folks from all over competing to solve crossword puzzles the fastest.

Quite possibly the most notable name in crossword puzzles is Will Shortz, who is the fourth editor of the *New York Times* crossword puzzle. Though his name appears on the puzzles, he is not always the creator.

As technology progresses and newspapers wane in popularity, the crossword puzzle has been brought to the digital world with web-based games and apps for smartphones, although nothing will ever compare to the feeling of being able to solve a crossword puzzle on paper with a pen. (It's hard! I was a puzzle editor for four years and I still can't do it.)

DID YOU KNOW?

Even though crossword puzzles were used in England to find decoders for the war effort, they were banned in Paris at the same time, out of fear they would be used as a means of relaying secret messages to the enemy.

TL;DR VERSION

- Arthur Wynne came up with the first crossword puzzle in 1913, based on a game he had played as a child called Magic Square.
- The *New York Times* harshly reviewed the crossword fad, but six years later began running the puzzles.
- A crossword puzzle competition was held in London as a way to find those skilled enough to crack Nazi codes.

THE HULA HOOP

This chapter is based on an episode of The Story Behind *podcast that was part of a series based on the song "We Didn't Start the Fire" by Billy Joel.*

While the origins of the hula hoop are unknown, Egyptian children played with dried vines in a hoop shape as far back as 3000 BC. They would either roll them around with sticks or twirl them around their waists. Greeks and Romans were also known to use the hoops as an exercise aid for their waists, although they used hoops made out of metal as well as wood and vines. Much of their art and pottery depict both kids and adult playing with hoops, including a vase that's now on display at the Louvre in Paris.

Hoops were also used by the Inuit (Eskimos) as targets to train children and adolescents to hone their hunting skills.

Rings have also played a role in religious rites and ceremonies for centuries (i.e., the wedding ring), and this included the hula hoop. American Indians used the hoop as part of their ceremonial dances, representing the circle of life. They used reeds to create the hoops, then bounced and swung them around their bodies to tell stories. In the 1930s, there was a revival of these dances by Tony White Cloud as a form of storytelling, and an annual Native American Hoop Dance competition is still held in Phoenix, Arizona.

Hoops were also popular throughout Europe over the centuries. In fact, they were so popular, medical records show a number of patients with dislocated backs and heart strain due to the hooping phenomenon.

But it wasn't until British sailors were introduced to hula dancers in the Polynesian islands in the eighteenth and nineteenth centuries that the term "hula hoop" arose. They noticed that the Hawaiian natives' swinging hip dances closely resembled the movement needed to keep hula hoops spinning around one's waist.

Hoop dancing became the exercise of choice in the late 1800s and 1900s, when Swiss composer Émile Jaques-Dalcroze created a special training for musicians and dancers using the hoop, which was said to express symmetry and spirit. It was called Eurhythmics. Annie Lennox learned about the exercise and named her music duo (with her romantic partner at the time, David A. Stewart) after it, although the spelling was changed to Eurythmics, without the *h*.

Hoops were not only toys for the waist; a popular game for children in Europe and colonial

America was to keep a hoop rolling by hitting it with a stick as they ran after it.

But the big craze for hula hoops was yet to come. In the 1950s, Richard Knerr and Arthur Melin developed a slingshot they used to train falcons and hawks by flinging meat into their beaks. They called their company Wham-O because of the sound the slingshot made.

When they heard about Australian children using swinging hoops made out of bamboo around their waists, they had the idea to use the newly available plastic to make a similar product in the US. They trademarked the name Hula Hoop and, by 1958, were manufacturing them.

They sold an estimated 25 million Hula Hoops in the first two months, and while it was a very short-lived trend, it was a big enough blip to get a mention in a verse of Billy Joel's "We Didn't Start the Fire," which is all about the trends and news stories of the 1950s.

The Hula Hoop continued to be a popular toy, although sales were never as huge as in the '50s. But it did have a little bit of a comeback as a new exercise craze in the 2000s, with notable names such as Michelle Obama, Catherine Zeta-Jones, and Kelly Osbourne using it for fitness. Queen Bey herself, Beyoncé, even brought hooping skills to the stage, performing her song "Work It Out." Classes in gyms and on beaches across the country began popping up, with participants using weighted hoops that not only added a bit of resistance to the workout, but also made it easy for those less skilled at the sport (ahem, that would be me) to keep up.

DID YOU KNOW?

Hula Hoops were once banned in Japan because gyrating hips were seen as indecent.

TL;DR VERSION

- Playing with hoops has been a pastime with both kids and adults for at least five thousand years.
- British sailors named the act of swinging a hoop around one's hips "hula hooping" after seeing Polynesians hula dancing.
- Wham-O was the first to trademark the name Hula Hoop and market plastic hoops as toys in the 1950s, which resulted in a short-lived craze.

HYPNOSIS

On the subject of hypnotism, you've no doubt heard the phrase "the power of suggestion." This power of suggestion can be traced back to religious ceremonies in ancient India, Greece, and Egypt. Back then, religion, science, and medicine were all one and the same. When someone got sick, it was common to be medicated by way of a ritual trance. This is considered by some to be the predecessor for what's known in modern times as hypnosis for pain management.

The precursor for hypnosis as we know it today, though, came from a German physician in the 1700s named Franz Mesmer. As you may have guessed, the word "mesmerize" came from his last name. He believed an invisible fluid that ran through the body was controlled by magnetism and, through putting his subjects into a trance-like state and using the power of suggestion, ailments could be healed through his control over the fluid.

While his magnetic fluid idea was proven false and he was ostracized, it was noted that many of those who had been participants in his parlor tricks did see an improvement in their health after being mesmerized.

In the mid-1800s, Scottish surgeon James Braid dismissed Mesmer's idea of magnetic fluid, but he did want to get to the bottom of the trance-like state and its physical effects. He was the one who coined the term "hypnosis," for the Greek deity Hypnos, the personification of sleep. Even when he realized this state of mind wasn't necessarily "sleep" and tried to change it, "hypnosis" had already become popular and stuck.

Even though most associate this state of mind with being relaxed and at ease, it was in a hypnotic state that clairvoyant Ella Salamon collapsed and died when she was asked to give medical advice about the hypnotist's brother. It was said at the time that she died because her brain couldn't handle the excitement. While we may never know the real reason for her death, since this was more than one hundred years ago, the Journal of the American Medical Association deems this the first person to die while under hypnosis.

Have you ever driven home and gotten so caught up in thinking of other things that you miss your exit? Or you've gotten home only to realize you don't remember the actual journey? These are common examples of times when we supposedly hypnotize ourselves.

Hypnotism, as performed by both therapists and performers, puts the subject into a state where they can eliminate external stimuli and concentrate only on what is being suggested to them.

Meditation and mindfulness techniques follow this same pattern. In fact, many videos and recordings used for hypnosis sound very similar to meditation prompts I've listened to before.

The point of all of it is to be able to use the power of suggestion for that part of the brain you don't realize you're using, the one that gives you muscle memory when typing at a keyboard or driving home without needing to pay attention to the signs to get there.

If a person can put suggestions into that area of the brain, it can become a habit, in the case of clinical hypnosis, or it can make you feel more comfortable and unaware of ridicule if you get called up to the stage during a hypnotist's performance and asked to dance up and down the aisles in front of two hundred students when you're in college.

Yup. That was me about fifteen years ago. And, even as I'm saying it now, I'm still not embarrassed that I pretended to know how to do ballet leaps. Maybe it was because one of the suggestions made while I was under hypnosis was that I wouldn't be embarrassed afterward by anything I did in the trance.

While hypnosis is recognized in some parts of the medical community as a valid form of therapy, others say it's a placebo effect, and we're only listening to the power of suggestion because of our need to please others and follow suggestions of those we hold in high regard.

As far as using it for pain management, neuroscientists have studied the brains of those with chronic pain under hypnosis and found a decrease in activity in the prefrontal cortex—the part that registers pain—when it's suggested that the pain is in fact minor. Hypnobirthing classes have become popular to help women in labor manage their pain, as well.

So, do those street hypnotists have the ability to walk into a delivery room in place of an epidural? Probably not. Those who seek hypnosis as a treatment for pain or even phobias or post-traumatic stress disorder have to work with a therapist over weeks or months of sessions for hypnotherapy.

But if you go to a hypnotism show, all audience members will be asked to try out an exercise for the performer to determine who is easily hypnotizable. About one in six people are what's known as somnambulists and can easily be hypnotized on the first try. (By the way, the word "somnambulist" comes from the Roman equivalent to the Greek god Hypnos, named Somnos.)

You're listening to one right now. I've been to two hypnotist shows in my life. And, both times, I was brought on stage and did things like forgetting what number comes after five and being made to feel that it was really hot or really cold within seconds and, yes, in my crowning achievement, performing ballet.

However, as those who have also been on stage will tell you, you don't lose control of yourself. You remember and are aware of everything you're doing.

Oh, and one more thing: The pocket-watch trick you normally see in pop-culture hypnotism is not a common practice, but sometimes an object like a pocket watch can be used to help focus the attention of an individual enough to help facilitate hypnosis.

DID YOU KNOW?

There are no nationally recognized regulations for hypnosis, although many therapists can be accredited hypnotherapists.

TL;DR VERSION

- The power of suggestion has been used since ancient times for pain management.
- Franz Mesmer believed that if a substance he called magnetic fluid ran through the body it could be useful in alleviating pain by putting people into trances.
- Those who are easily hypnotizable are known as somnambulists. This ability doesn't necessarily correlate with that person's level of intelligence.

PING-PONG

This chapter is based on an episode of The Story Behind *podcast that was part of a series based on pop-culture references in the movie* Forrest Gump.

Back in the late 1800s, lawn tennis was a popular game among the elite in Victorian England, India, and South Africa. When people wanted to bring the game indoors for the winter or in inclement weather, they improvised equipment by using dining tables and stacks of books for nets, rubber balls or rounded champagne cork tops for balls, and books or cigar-box tops as paddles.

What we now know as ping-pong had a number of names, including Whiff-Whaff, gossima, Flim Flam and Pim Pam. If you're looking for ping-pong as an Olympic sport, you won't find it listed—you'll see it referred to as table tennis because the name ping-pong is actually trademarked by Parker Brothers, but it ended up becoming so generic, you might not realize it, like the words "Dumpster" or "Crock-Pot."

The Ping-Pong Association even had to change its name to the Table Tennis Association in 1922.

When sandwich rubber was invented in 1950, it was a game changer—literally. Sandwich rubber is what you see on a table tennis paddle, which is sponge with a layer of rubber over it. When the Japanese introduced this to the game, players were able to put spin on the ball and hit it in ways they couldn't before. The International Table Tennis Federation quickly added new regulations concerning the use of the new paddles in play.

The rules and regulations needed to be changed again after 2000, when more games were being televised. In order to attract a larger audience and keep their attention, the twenty-one-point scoring system was brought down to eleven, and the size of the ball was increased for better television visibility. The greater size of the ball also increased its weight, which slowed down the game for visibility as well. This was before we all had flat-screen, hi-def televisions, by the way.

Believe it or not, there are an estimated forty million competitive table tennis players in the world, and even more millions who play it recreationally—I don't know if this number includes beer pong, but it is interesting to note that beer pong was invented in 1950 at a Dartmouth College fraternity, which originally involved paddles and later evolved into the drinking game most college students and adults know today. Ping-pong, though, is probably most popular in China, which you may remember being referenced in the book and movie *Forrest Gump.*

When we hear about the Cold War, many probably think it was between America and Soviet Russia, but in addition to what we called the Iron Curtain, separating the Soviet bloc from the western world, there was another curtain, known as the Bamboo Curtain, that separated Communist China from the United States following Mao Zedong's 1949 revolution.

More than twenty years later, China's ties with the USSR were beginning to fray, and Chairman Mao began to look for ways to open up communication with the United States again. This occurred at the time when President Richard Nixon had made relations with China one of his top priorities.

When Glenn Cowan, a nineteen-year-old professional US ping-pong player, boarded a shuttle bus during the 1971 World Table Tennis Championship, Zhuang Zedong, a player from the Chinese national team who was also on the shuttle, stepped up to him and shook his hand, speaking to him through an interpreter, and presented Cowan with a silk-screened picture of China's Huangshan mountains. The next day, Cowan presented Zedong with a T-shirt with a peace symbol and the Beatles' lyrics to "Let It Be" on it.

Originally, the Chinese players were told not to speak to the Americans, but after the story and photographs of the exchange circulated in newspapers in China and the US,

Chairman Mao invited the US team to take an all-expense-paid trip to play ping-pong in China, which was referred to in *Forrest Gump* and became known as "ping-pong diplomacy."

China was vastly different from anything the US team had seen before, and the stigma relating to Americans hadn't fully worn off by the time they arrived. Even the banners welcoming the team were hung over graffiti that said, "Down with the Yankee Oppressors and their Running Dogs."

But political leaders made sure the Americans were treated as welcome guests. At a banquet in April of 1971, Premier Zhou En-lai told the team and the press, "You have opened a new chapter in the relations of the American and Chinese people. I am confident that this beginning again of our friendship will certainly meet with majority support of our two peoples."

The next day, the twenty-year trade embargo with China was lifted by the United States.

In the movie, Forrest Gump was depicted as being so good at ping-pong that those in the hospital with him were entranced by his playing—so much so that they are shown surrounding him at a ping-pong table, instead of watching a television showing Apollo 11 landing on the moon in 1969.

When Forrest is taught to play ping-pong, he is told to never take his eyes off the ball. If you watch Tom Hanks in these scenes, you'll notice he doesn't even blink. This may be easier for a seasoned player to do, but it's a natural reflex to blink sometimes if an object is coming toward your face.

Was Tom Hanks just that good? Well, yes, because he's Tom Hanks, but also because many of his ping-pong scenes were shot without a ball, including the intense match he has in China—the ping-pong ball was added in later with the help of computer-generated imagery. Hanks and his Chinese opponent, listed in the credits as Valentine, listened to a metronome and timed their swats to the clicks.

The movie was known for its innovative use of computer graphics, like seamlessly inserting Tom Hanks into historical footage and removing Gary Sinise's legs for his character, Lieutenant Dan. Even the floating feather seen at the beginning and end of the movie was computer-generated. It's no wonder the movie took the Oscar for Best Visual Effects in 1995 as one of its six that year.

DID YOU KNOW?

Pop singer Justin Bieber loves table tennis so much, he brings a table and equipment on tour with him so he can set up a game on-the-go wherever he is.

TL;DR VERSION

- Ping-Pong is trademarked by Parker Brothers.
- Table tennis was originally a version of outdoor tennis brought inside during Victorian times with the use of dining tables, books, and champagne corks as equipment.
- The game of ping-pong helped end the trade embargo between the United States and China.

ROLLER SKATES

If you're worried I'm going to go into a big gush-fest over Olivia Newton-John and Gene Kelly in *Xanadu* for this chapter, don't worry. I'll save it for the end, since the movie was released more than two centuries after roller skates first wheeled into the public eye.

Most historians and the National Museum of Roller Skating in Lincoln, Nebraska, point to John Joseph Merlin as the original inventor of roller skates. He was known for his quirkiness and outlandish behavior, including making an entrance to a hoity-toity masquerade party in London during the 1760s wearing his skates and playing the violin.

Unfortunately, his skates didn't include a braking mechanism and he wasn't very good at steering. As he entered the party, he crashed right into a giant mirror in the ballroom, ending not only the party but his fascination with skates. He continued to tinker with them, but without any way of braking, they didn't catch on, and he never patented the idea.

The early versions of skates were more like the inline skates we're used to seeing today. Merlin's skates consisted of two wheels on each boot, while the follow up patented by Monsieur Petitbled of France in 1819 consisted of two to four wheels made of metal, ivory or wood, configured in a straight line on a metal plate that could be attached to a boot. The year before, roller skates were used in the German opera *Der Maler oder die Wintervergn Ugungen* to mimic ice skating.

Various inventors tried different configurations of wheels to make turning easier for the skater, but it was New Yorker Leonard Plimpton who came up with the idea of the quad skate in 1863, using two wheels in front and two in the back. The wheels were made of rubber and put on pivots to allow for skaters to turn much more easily than before, and the skates could be fastened to the wearer's regular shoes with leather straps.

He used the floor of his New York City furniture store as a makeshift rink and leased his skates out for others to enjoy. He also founded the New York Roller Skating Association, which opened the first public skating rink in the United States in a renovated dining room in a hotel in Rhode Island in 1866.

Plimpton's design became instantly popular. With new labor laws emerging, shortening the work day and giving people more leisure time, roller rinks began opening up for kids and adults alike to enjoy the new hobby.

In one of its many dips in popularity, World War I and the Depression made roller skating less affordable. But the trend soon picked back up when streets were paved and sidewalks laid. The trend continued for years with moderate success, but probably the most well-known spike in roller skating popularity came in the 1970s.

The advent of the disco movement is said to be the cause of a new love of roller skating. It was promoted as good exercise, which many celebrities were seen enjoying. Pop culture also pushed the trend, with songs, television specials, and movies all about roller skating, including *Roller Disco* and even Patrick Swayze's acting debut in *Skatetown, USA*. Oh! And quite possibly one of the greatest movie musicals ever, *Xanadu*!

Skating crashed again when the '80s brought a hatred for disco and, as with the Depression earlier, economic turmoil resulted in many skating rinks closing their doors. However, right before the turn of the decade, two brothers, Scott and Brennan Olson, came across a pair of antique skates with the old inline wheels. Looking for a way to practice their hockey skills off the ice, they attached polyurethane wheels to ice hockey boots. They trademarked the name Rollerblade, and, by 1983, were incorporated.

By the '90s, skating rinks that had avoided shutting down in the previous decade and newly built ones began seeing an uptick in adolescents rollerblading.

Nowadays, anything that induces nostalgia can be considered trendy, including a few more roller-skating rinks opening up, despite the rising cost of the real estate needed to erect buildings big enough. One thing I should mention is that, as long as roller skates have been around, so have roller skating races, and conjointly, roller derby. By the way, if you're as much of a fan of puns and wordplay as I am, you should check out some of the names used by roller derby teams and participants, like the Stop Drop and Rollers.

DID YOU KNOW?

Carhop service at restaurants, with the waitstaff delivering food to cars on roller skates, was common in the 1950s and '60s. The trend is still embraced by Sonic, which also features an annual Skate-Off between all 3,500 restaurants.

TL;DR VERSION

- John Joseph Merlin is known as the first to publicly display roller skates, although he ended up breaking a very large and expensive mirror in the process.
- Leonard Plimpton designed the quad-style roller skates with rubber wheels, making turning much easier than with past designs.
- *Xanadu* is a much-underappreciated movie and came out at the height of roller skating fever.

THE SLINKY

Remember in grade school learning that Isaac Newton came up with the theory of gravity when an apple fell from a tree? Well, a similar inspirational moment happened to Richard James back in 1943.

Newton was on an academic break caused by the Plague when he was drinking in his orchard and, as apples were falling, he began to wonder why they fell straight down, instead of sideways. This became the beginnings of his theory.

However, James wasn't quite on a break from work when he came up with his concept of a springy toy. At the time, James was working at a shipyard, and was in the process of developing a spring that could help keep objects on vessels secure out at sea. The story goes that he accidentally knocked a spring to the floor, which, instead of landing on its side, landed on its end and flopped over and over itself in a "walking" motion.

We're all probably familiar with Shiny Object Syndrome, right? Well, this was more like Shiny-and-Springy Object

Syndrome for James, who spend the next two years developing this as a toy. He began experimenting with different materials and lengths and developed a machine that wound 80 feet of wire into the perfect springy toy.

His wife, Betty, was totally on board with her husband's invention, and even came up with the name "Slinky," which she picked from the dictionary when she read the definition was "graceful and sinuous in movement."

With the help of a $500 loan, James was able to manufacture the toys and sell them to stores in 1945, but sales were slow at first. No one quite knew what to do with a spring for a toy. It wasn't until Richard and Betty did an in-store demonstration of the Slinky at Gimbel's Department Store in Philadelphia that sales picked up. In fact, the 400 the couple brought with them sold out in 90 minutes once they showed off what it could do.

Betty not only played an integral role in naming the new toy and early marketing, but she is also behind the company's later success. While Richard grew to enjoy the spotlight and success of the Slinky, he began donating large sums of money to religious organizations and eventually gave up his company, James Industries, and moved to Bolivia to join what some say was a cult.

But Betty stayed behind with the couple's six children and took over the company. However, sales began to stagnate by the 1950s—another reason Richard was said to have left after leaving the company on the verge of bankruptcy and accumulating unpaid bills because of his donations of funds to his religious groups.

Betty turned that financial slump upside-down when she moved the company to her hometown of Hollidaysburg, Pennsylvania, and began promoting the Slinky through catchy advertisements on TV, with slogans still known today, like "What walks down stairs, alone or in pairs, and makes a slinkity sound?" (If the song is now stuck in your head, you're welcome!)

The Slinky also owes some of its upturned luck to Slinky Dog, invented by Helen H. Malstead when the original Slinky was given to her son at Christmas and he wondered what would happen if it were put on wheels. Malstead came up with the idea to make a pull-toy out of the spring and sent Richard sketches. Before he departed for Bolivia, they worked together to create Slinky Dog, and its success (along with Betty's marketing) helped turn the business around after he leFort Slinky Dog's popularity was renewed with the inclusion of the character in 1995's *Toy Story*, voiced by Jim Varney, who was known for his portrayal of Ernest in the children's television show and movies.

The Slinky was named the official toy of Pennsylvania in 2001. When Betty passed away in 2008, the *New York Times* reported that more than 300 million had been sold, and, if linked together and stretched, would circle the earth 150 times.

DID YOU KNOW?

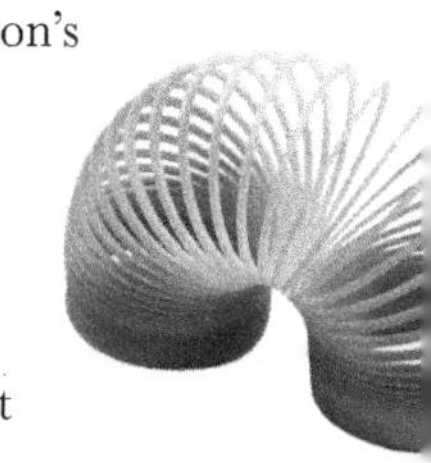

Not so much related to Slinky, but more about Newton's apple anecdote we all heard growing up: There's no evidence to suggest the apple fell on his head, as it's usually depicted in animations and children's books. By the way, the original apple tree that inspired Newton's idea is still growing and is a tourist attraction in the UK.

TL;DR VERSION

- Robert James came up with the idea of the Slinky when a tension spring he was working on fell off his work table and rolled over and over itself.
- His wife, Betty, came up with the name and also took over the company and pulled it out of financial troubles when Robert left for Bolivia out of the blue.
- The Slinky is the official toy of the state of Pennsylvania.

THE TREADMILL

When you think about the actual purpose of a treadmill, it's easy to see why some European countries would scoff at the idea of a stationary sidewalk, as many do now. But the idea for the treadmill began in the late first century in Rome, when the treadmill was introduced as a sort of human-powered crane to lift heavy objects. Romans would actually walk within the diameter of a large wheel, kind of like a human hamster wheel. As funny as that sounds, they were able to lift more weight with fewer crew members using this system.

Using a treadmill for power continued to be a common occurrence through the centuries. In fact, the term "horsepower" comes from the measurement of strength of a horse needed to operate a treadmill or similar stationary machine as a source of power for farms in the 1800s.

In 1818, a British engineer by the name of Sir William Cubitt invented the prison treadmill, also known as a "tread-wheel." Convicts were made to step on spokes of a large paddle wheel to generate power for tasks such as grinding corn and pumping water; however, there are reports of prisons using treadmills as a form of punishment, rather than to produce any sort of power. Use of the treadmill for prisoners was finally deemed too cruel in England when the Prisons Act of 1889 abolished hard labor for convicts.

You would think that would have been the end of treadmills, but it was only a matter of time until the treadmill made a reappearance in the form of exercise equipment. A prisoner from back then would probably be shocked to see people paying $20 a month just to do such a thing nowadays.

The treadmill wasn't quite ready to make the leap to recreational exercise equipment in 1952, but it did become part of the medical community when Dr. Robert Bruce and his colleague, Wayne Quinton, used a treadmill as a way to monitor heart conditions and diseases. He developed the Bruce Protocol diagnostic test, which is still used to evaluate cardiac function today, and he's known as the father of exercise cardiology.

When you hear the word "aerobics," you may get a mental picture of Olivia Newton-John or Jane Fonda in Spandex and leg warmers in the '80s. But that wasn't the beginning of the aerobics craze.

Dr. Kenneth H. Cooper developed aerobics at an Air Force Hospital in the 1960s with the intention of preventing coronary artery disease in members of the military, but it took off as a fun way to stay in shape.

With aerobics, Cooper began popularizing the treadmill, but he would also go on to use it as part of a test of cardiovascular fitness, otherwise known as the treadmill stress test. If you've ever seen a television show in which someone is having their physical fitness tested by having monitors attached while being asked to run on the treadmill, you're most likely seeing Cooper's method.

By the way, Cooper is also behind the idea of running for fun (whatever that means).

It seems that, every year, new exercise equipment is introduced to the fitness scene, yet treadmills continue to be the No. 1-selling cardiovascular exercise machine. Go to any gym, especially in the beginning of the year, to see just how popular these machines are.

Regardless of infomercials promising us a better, more efficient way to get fit, walking remains one of the easiest ways to incorporate movement into our routines. Many people can even multitask while walking on the treadmill.

Another reason treadmills are popular, as opposed to just going outside and taking a walk, is this wonderful technology known as climate control indoors. Instead of trudging in the snow or melting in the heat of summer, you can walk as long as you want indoors on a treadmill without worrying about slipping on ice or getting a sunburn.

Plus, you're never far from where you started when you take a walk on a treadmill.

DID YOU KNOW?

If you have a treadmill that's basically used as an expensive laundry rack, you're not alone. Forty percent of people who buy home exercise equipment say they ended up using it less than they originally expected.

TL;DR VERSION

- The treadmill started as a human-powered crane in first-century Rome.
- Treadmills were once used in prisons as a way to generate power, but also as a form of punishment.
- Aerobics was developed in the 1960s as a way to prevent coronary artery disease for those in the military.

WATER GUNS

Believe it or not, there's not a lot known about water guns before they were patented in 1896, but the idea is much older than that. The pastime of having pretend battles with water-filled weaponry had been around for decades, but those water guns weren't all-inclusive devices like ours today. They were made up of a pouch filled with water, attached to a cast-iron gun casing with a tube. Instead of a bullet chamber, the water was shot through the hole in the front as someone simultaneously squeezed the pouch to pressurize the water.

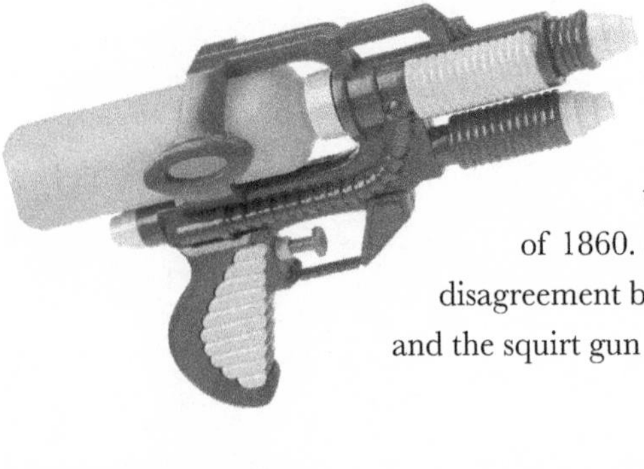

On hot days, this was a welcome relief, but also a form of mischief, with some going great lengths for their squirt guns. Back in 1858, the sophomore class of 1861 at Amherst College decided that the tradition of hazing freshmen with a water gun was outdated and went all out preparing an elaborate funeral service for the squirt-gun procured by the junior class of 1860. But there was a disagreement between the classes, and the squirt gun was stolen by the sophomores, and that was the same squirt gun they intended to bury at the ceremony, titled "Funeral Services and Wake at the Burial of S. Gunn, X-Member of the Class of '60."

While the sophomores ate their supper before the ceremony, the juniors plotted to stop the haughtily-named funeral, and soon the lone sophomore guard's cry of "'61! '61!" rallied their side as the juniors charged toward the building where the squirt gun was being held, in a coffin, of course. The sophomore guard, holding an actual pistol to guard a coffin with a squirt gun in it (I know!), managed to hold off the juniors until reinforcements arrived.

Unfortunately, the building, known as the *Ultima Thul*e, took the brunt of the battle between the two classes, with wood from the stairs and doors broken off and used for weapons. Over a squirt gun.

Finally, the president of Amherst arrived on the scene to stop the fighting, siding with the sophomores for their good intentions, and the funeral (for a squirt gun, let me once again remind you) took place, complete with lemonade, a peace pipe, and a wake. But the fight went down in Amherst legend and became known as the Squirt-Gun Riot of 1858.

A few changes to the squirt gun were made over the course of several decades: A trigger was added to squeeze the pouch, plastic soon replaced the cast iron used for the squirt guns, and soon even the pouch was taken away and water could be stored in the squirt gun's handle, so it could be used like a common household spray bottle.

In 1977, a motorized pressure tank was added so that, instead of producing a squirt of water every time the trigger was pressed, the user could hold down the trigger and shoot a stream of water. However, the batteries to run the motor were expensive, and the toy didn't gain a lot of traction.

In 1989, Lonnie Johnson, a mechanical and nuclear engineer, was working on a heat pump for the Galileo Jupiter mission. When he took it home and began tinkering with it, he found it could shoot water farther and faster than the squirt guns then on the market, and he realized it would make a great toy for kids. He developed the hand-pumping motion to pressurize the water stream without the use of batteries, but it would take a few years to fully develop the first Super Soaker.

The first company Johnson went to with his design had it on a waiting list for two years; the second went out of business because squirt-gun sales were waning due to their similarity in appearance to real guns. But the third, Larami, was able to manufacture the squirt gun and propel it to the top of the toy charts in the early 1990s.

One of the most popular features was the large water-bottle attachment to the top of the gun that meant kids didn't have to constantly refill the supply.

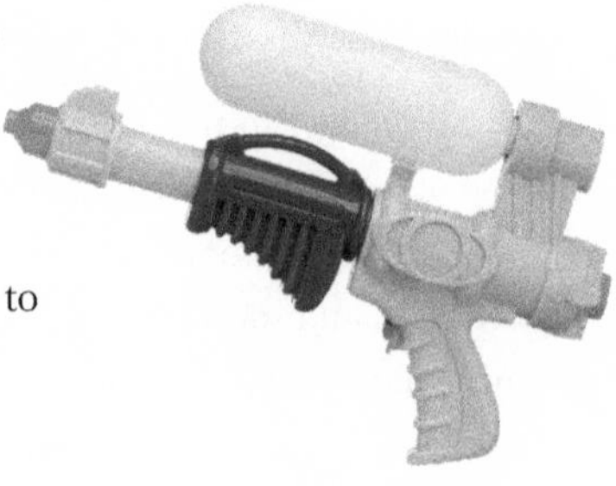

In an almost no-brainer move, Super Soaker was merged with Nerf in 2010 and from there, the company continues to grow, with new products being added to its line every summer.

DID YOU KNOW?

When Abraham Lincoln began calling for volunteers to fight in the Civil War, Union General Sherman thought Lincoln had underestimated how many soldiers were needed, and he's quoted as saying, "Why, you might as well attempt to put out the flames of a burning house with a squirt gun."

TL;DR VERSION

- Water guns were originally metal gun frames fitted with a tube for water to flow through from a pouch that the owner had to squeeze to pressurize the water.
- In 1858, a riot occurred at Amherst College between sophomores and juniors over the Squirt Gun used to haze incoming freshmen. Go back and read about the riot, it's ridiculous.
- Lonnie Johnson was working on the Galileo Jupiter mission when he had the idea for the Super Soaker.

PART FIVE:
IN THE HOME

BAND-AIDS

As far back as 1500 BC, Egyptians used honey to treat wounds and cover cuts and scrapes. Gauze became more popular in later centuries, especially in the 1860s, when physician Joseph Lister added sterilized gauze to his ongoing program for antiseptic surgery and care.

In 1920, Josephine Dickson was prone to frequent minor cuts and burns around the house. The choices she had for tending to these wounds weren't very easy to self-administer. If it were an infomercial, you would see someone in black and white leaving the cuts open, risking infection, or trying to wrap a piece of gauze around her fingers and tying it with just one hand, or attempting to construct a makeshift bandage.

In that infomercial, someone would inevitably say, "There's got to be a better way!"

It happened that her husband, Earle Dickson, was a cotton buyer for Johnson & Johnson and had access to surgical tape and bandages. With the addition of a stiff fabric known as crinoline, used in petticoats, to keep the gauze sanitary, the two of them stuck cut pieces of bandages onto pre-cut strips of surgical tape to come up with the prototype for what we know today as the Band-Aid.

By the way, Band-Aid is the brand name. Well, actually, BAND-AID® Brand Adhesive Bandages is the formal name.

You would think Band-Aids would have taken off immediately, but they were slow to catch on. It wasn't until Johnson & Johnson supplied them free to none other than the Boy Scouts of America for their first-aid kits that they finally caught on. Boy Scouts carried them in a cardboard package attached to their belts and used Band-Aids as a way to earn their merit badges.

When other companies tried to crowd the adhesive bandage market in the 1950s, Johnson & Johnson were quick to make sure their product would blow the others out of the water by making sure it was the stickiest. They showed this feature in a commercial where Band-Aids would stick to an egg, even in boiling water.

One thing Johnson & Johnson was a little blind to was the fact that they touted their bandages as "flesh-colored" in that commercial, when, back then, they only came in one color—a pinkish beige.

Michael Panayiotis had the inspiration to create adhesive bandages in other flesh tones to cater to the black and Hispanic populations. He called this product Ebon-Aides, and boxes would come with adhesive bandages in shades called honey beige, cinnamon, coffee brown, and black licorice.

Unfortunately, when stores bought these bandages, most would place them on shelves with products specifically for black people instead of with the other boxes of adhesive bandages, and they were a bust, save for a few people who still reach out to Panayiotis for any he may still have available.

Johnson & Johnson have since started making clear bandages in an effort to help them blend with any skin tone.

The brand name became a clever pun on two occasions.

The first was back in 1984 with Band-Aid, one of the biggest collaborations of musicians ever, who got together to perform the Christmas radio standard, "Do They Know It's Christmas?" In November of that year, Bob Geldof of the Boomtown Rats and Midge Ure of Ultravox collaborated to write a song after seeing reports of starvation and famine in Ethiopia, contacting some of the best known musicians of England and Ireland to record it. The idea was to make a song so big as to call attention to the crisis, and also to use the profits from the sale of the record to help the cause.

They only had twenty-four hours of studio time to record, mix, and release the song, and somehow, they got more than thirty musicians and bands on board to do so, including Sting, George Michael, Bono, Phil Collins, Duran Duran, and Spandau Ballet. One of the biggest names scheduled to record was Boy George, who slept late, missing his Concorde from New York to London, was woken up by Geldof, and rushed onto another Concorde to get to the studio at the last minute to record his vocals for the song.

But it was worth it. The supergroup raised $8 million with the release of the record. Even when it looked like the British Government was going to impose taxes on the single, Geldof stood up to then-Prime Minister Margaret Thatcher, who still collected the tax, known as VAT, but then donated the revenue from that tax to the relief efforts.

The second time the brand name was used in pop culture as a clever pun was in the 2000 Cameron Crowe movie *Almost Famous*, when Kate Hudson's character Penny Lane called herself and her friends "Band-Aids," as opposed to groupies, of the band Stillwater.

DID YOU KNOW?

The term "gauze" appeared in the sixteenth century, but etymologists are uncertain about its origins. It's been said there's a connection to the Palestinian city Gaza, but no evidence exists to support that.

TL;DR VERSION

- The first known wound covering was honey, used by Egyptians as far as back 1500 BC.
- Josephine Dickson, the wife of a cotton buyer at Johnson & Johnson, was prone to accidents around

the house. She and her husband combined gauze, crinoline, and surgical tape to create a bandage that was self-sticking and easy for one person to self-apply.

- The members of Band-Aid only had twenty-four hours in the studio to record, mix, and master "Do They Know It's Christmas?"

DIAPERS

Diapers have come a long way since their first known usage in ancient times, although what they were using them for would be unheard-of now. Just imagine a commercial on daytime TV touting the absorbency of milkweed leaves, or of animal skin filled with grass or moss. In warmer climates, it was common for babies to be left naked and cleaned up after.

As long as there have been humans, there has been baby poop, and, if you're a parent, you know how crucial always having clean diapers in the house can be, especially for those 2:00 a.m. wake-ups.

Now, for the most part, we've moved away from using animal skin on our babies' bums, and have been using cloth for hundreds of years. By the late 1800s, most babies were wearing linen or flannel that was folded and pinned, but, while we now change our babies' diapers every few hours, it was common back then to leave the diaper on for days, until the diaper sagged and couldn't hold anymore.

Luckily, by the turn of the century, the western world became more aware of the importance of hygiene and began changing diapers more frequently. Consistent with the roles at the turn of the twentieth century, women were mostly in charge of child-rearing duties, such as changing and laundering the soiled diapers.

A type of disposable diaper insert was invented in Sweden in the 1930s, made from paper and a softwood pulp, but was seen as mostly a product for families that frequently traveled. More absorbent materials were being developed at the same time. It was only a matter of time before this innovation, plus the Baby Boom following World War II, would give rise to more convenient diapers.

When the war broke out, many women had to go to work to support their families and fill in positions while men were away at war. This meant having less time for washing all the dirty cloth diapers, which led to diaper delivery services becoming popular.

After the war, a housewife and mother of two named Marion Donovan was fed up with constantly changing not only the soiled cloth diapers, but also bed sheets and clothing. Parents of that era had to choose between cloth diapers and the mess they involved, or rubber pants, which were known for causing diaper rash and pinching the baby's skin.

Donovan decided to create a waterproof diaper cover, which she sewed from a shower curtain and, later, parachute nylon. She included in her design and patents the use of plastic snaps, which would take the place of the diaper pins that were known to pop open and prick either the parent or baby during changing.

She named her invention a Boater and began trying to sell the design to manufacturers, but was consistently turned down, with the manufacturers telling her they were not receiving requests for something like Donovan's Boater, and were perfectly happy with what was already on the market.

Donovan didn't give up, and she began manufacturing them on her own. She was able to sell her Boaters to Saks Fifth Avenue and soon sold out.

The Boater still required the parent to wash it in order to reuse it, and Donovan had the idea to create a disposable diaper out of paper. But, again, she couldn't convince manufacturers that it would be a product worth making and selling.

In 1948, Johnson & Johnson introduced the first alternative to cloth diapers, imported from Sweden and called Chux. The one-piece system was mainly advertised as being for traveling, but the ease of disposal made it popular in Europe and the United States.

Victor Mills was a chemical engineer at Procter & Gamble in the mid-'50s when he began to investigate the idea of a truly disposable diaper that could easily incorporate the company's recent acquisition of Charmin Paper Mills, but the diaper failed initial market testing when parents didn't like the amount of plastic in it, especially in hot weather.

Once Procter & Gamble developed a diaper with less plastic they called Pampers, sales were slow. Not only did these diapers require safety pins to keep them in place, but they were also more expensive than cloth diapers. Once they were able to lower the cost of manufacturing and streamline operations to lower the price, though, Pampers became the company's second-largest brand.

Other diaper companies quickly followed suit, including Johnson & Johnson dropping their Chux line in favor of their own brand of disposables, but, by 1981, the disposable diaper industry was too crowded and Johnson & Johnson backed out.

The addition of Kimberly-Clark's Huggies brand had also taken the market by storm in 1978, when they introduced an elastic waistband and leg band, but Procter & Gamble filed a lawsuit against Kimberly-Clark, contending they had infringed a patent for an elastic used in the waistband, which was countered with an antitrust suit against Procter & Gamble.

These diaper wars were finally settled in 1992, with Kimberly-Clark winning and also becoming the No. 1 brand in America. Procter & Gamble had the market share, but when they introduced a less expensive option, Luvs, they split their customer base in two, leaving both brands to battle for No. 2. (Yes, that pun was absolutely intentional.)

That wasn't the only instance in which people disagreed over diapers.

Dr. T. Berry Brazelton, a pediatrician who was concerned at the number of cases of bedwetting he was seeing in the 1960s, began to advise the mothers of his patients to let the children decide when they were ready to quit their diapers for good and begin toilet training.

The successful results of this study happened to coincide with Pampers introducing bigger diapers for older children not yet potty-trained. The company even took on Brazelton as a paid consultant. But other pediatricians have framed this as a conflict of interest and have denounced the later toilet training methods Brazelton recommended.

Today, just to add yet another thing to exacerbate the constant second-guessing parents put up with when they have a baby, the reusable diaper trend has come back full-force, giving parents an eco-friendly option.

DID YOU KNOW?

In 1840, the safety pin was invented, which became the go-to method for keeping diapers on babies.

TL;DR VERSION

- Children in ancient times were wrapped in leaves and animal skins with moss or grass stuffed in them as diapers.

- Marion Donovan was an innovator when it came to creating a diaper that was effective and easy to use, but manufacturers turned her down.
- The doctor who promoted later toilet training was paid to be a consultant for Pampers, leading many doctors today to question the integrity of his studies.

THE DISHWASHER

Some of the earliest artifacts found denoting human life in ancient times have been pottery and cookware, and one of the reasons nomads began to settle in the Fertile Crescent was the availability of water nearby. Not only was this useful for drinking and crop irrigation, but water was necessary for cleaning those pots and dishes.

Early civilizations had to carry their dishes to the local water supply, or carry water back to their homes to hand-wash their cookware. Luckily, as the years went on, waterways became more common, as mentioned in a few other chapters in this book ("The Story Behind the Lead Pipe," "The Story Behind the Fire Hydrant"). Washing dishes became easier with water in the home becoming more accessible.

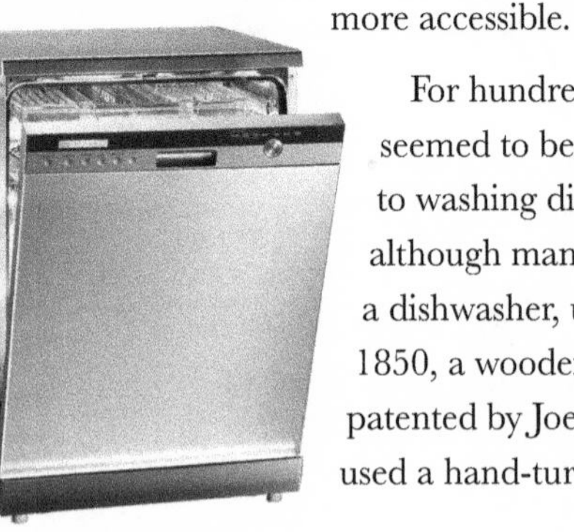

For hundreds of years, there seemed to be no alternative to washing dishes by hand, although many tried to invent a dishwasher, unsuccessfully. In 1850, a wooden machine was patented by Joel Houghton which used a hand-turned wheel, but it barely did the job without more manual labor than simply hand-washing the dishes.

But those with the means to do so were able to hire servants to do their dishes for them.

Josephine Garis Cochrane was in one of those upper echelons, but found herself constantly annoyed with her servants' clumsily chipping her fine china while washing it. She began washing the supposedly two-hundred-year-old dishes by hand herself.

Her father had been an engineer and her great-grandfather was an innovator in the world of steamboats. This was also during the Industrial Revolution, when any task seemed like it could be automated by a machine. So she set out to create a machine that could wash dishes more efficiently than human hands.

One story recounts that, on the morning when she had this idea, she stopped in the midst of dish-washing to start researching in her library, not even realizing she was still holding a cup in her hands. She soon came up with the idea of a rack that would hold dishes in place while water pressure washed and rinsed them.

She began measuring her plates and cups and designing wire compartments to hold them, which would then be placed in a copper boiler. The dishes would turn on a wheel while hot soapy water was used to clean them. Following her husband's death soon after, she realized she needed her dishwasher to be a success not only for her own convenience, but also to settle the many debts he left behind.

After showing her design to a few engineers, she became frustrated when they began to tinker with it, only to realize her original design was better than anything they could come up with.

"They knew I knew nothing, academically, about mechanics, and they insisted on having their own way with my invention until they convinced themselves my way was the better, no matter how I had arrived at it," she's quoted as saying.

Finally, she worked with mechanic George Butters to bring her design to fruition, and filed the patent in 1886.

Unfortunately, the success she was hoping for with housewives didn't seem to be there from the beginning. Nevertheless, she showed her invention at the 1893 World's Columbian Exposition, and hotels and restaurants had enough interest to make it the success she was hoping for. Her company, the Garis-Cochran Manufacturing Company, became known as KitchenAid and is now a part of Whirlpool.

Dishwashers haven't waned in popularity, although it took a few decades for them to become more readily available for households, but within decades became common (in all sorts of garish colors the '70s had to offer). Dishwashers can be bulky, taking up valuable space in smaller kitchens (we actually have one that sits on the countertop in our tiny kitchen), so hand-washing is still a necessity for many.

DID YOU KNOW?

Ever leave dishes in the sink to soak (or have a lazy household member do this with no intention of actually washing the dish in question)? Well, it's actually one of the worst things you can do, hygienically speaking. Soaking dishes in water that's cooler than sixty degrees Fahrenheit is actually a breeding ground for bacteria. So, no more excuses!

TL;DR VERSION

- If you think washing dishes is annoying now, imagine having to bring those dishes to the local water source to clean them, or having to carry back enough water to wash them at home.
- Wealthy socialite Josephine Garis Cochrane was tired of her servants chipping her heirloom china after she entertained guests, so she decided to create a machine to wash her dishes instead.
- Once she was able to create her dishwasher, hotels and restaurants became interested and bought her first units.

THE LEAD PIPE

This chapter is based on an episode of The Story Behind *podcast that was part of a series based on the weapons from the game Clue.*

Pipes and plumbing have been around since at least 3000 BC, according to archaeologists who've found copper pipes in palace ruins in India, and the first known flush toilet was in Crete in the 1700s BC. It's funny to think of lead as being an *improvement* to plumbing, especially because of what we know now about lead poisoning and the problems in places like Flint, Michigan. But the Roman Empire's system for bringing water into Rome, known as the aqueducts, used not only lead for sanitary reasons, but also bronze and marble from 500 BC to the mid-fifth century AD.

Lead has been a popular metal for western civilization for centuries. If you've ever wondered why the element is portrayed as Pb on the periodic table of elements, that comes from the Latin word for it, *plumbum*. (Yup. *Plumbum.*)

So, did the lead pipes affect the health of the Romans? Well, Vitruvius, an architect known for his work under Caesar and Augustus and probably most recognized as the man portrayed in Leonardo da

Vinci's drawing, once remarked, "Water conducted through earthen pipes is more wholesome than that through lead; indeed that conveyed in lead must be injurious, because from it white lead is obtained, and this is said to be injurious to the human system."

Now, there were a lot of implausible statements made in those days about health, but this ended up not being one of them, given that he had observed the poor health of those working with lead. Lead was not only used for waterworks, but also for making cookware like pots and kettles.

It's even been suggested that lead poisoning was one of the contributing factors to the fall of Rome. It's actually highly unlikely, considering they constantly kept the water running through the lead pipes, and mineral deposits from their hard water would coat the pipes, creating a buffer between the lead and the water. Nowadays, we shut off water in our homes for the most part, leaving water to sit in the pipes, and if those pipes are made of lead, there's a greater chance that deposits will find their way into the water.

When we think about the lead pipe in terms of its use as a weapon in the game of Clue and its subsequent movie version, it could take on a whole new meaning.

Would it be possible for Mr. Boddy to die from a lead pipe?

Maybe from the whack to the head, but indulge me in a bit of a rabbit hole, will you?

Lee Ving, lead singer of the punk band Fear, played Mr. Boddy in the movie Clue. He was thirty-five when the movie came out. Let's say Mr. Boddy was the same age as Ving in the year in which the movie is set, 1954. That means he would have been born before the 1920s.

At that time, lead was everywhere. His mother (presumably Mrs. Boddy) probably ingested lead into her system when she was pregnant with Baby Boddy.

The toys he played with most likely were painted with lead paint, not to mention that his room and house were painted with lead paint, as well. There's a chance he could have been served food on plates and in bowls decorated with lead glazes, too.

As a rich man, he would have maybe owned a car or two, which ran on leaded gasoline, and cars continued to do so into the 1970s, which is why you see gas labeled as "Unleaded" on your pump today.

Even though it was widely known to be toxic as early as the previous century, the lead industry was booming. By the 1950s, however, public health officials were making the dangers of lead in the home known, and children were brought to doctors to be tested for the levels of lead in their system, which were often elevated and linked to the water they were drinking and to paint in their homes.

So, even though we may not know if Mr. Boddy died from a lead-pipe whack on the head, I can guarantee Mr. Boddy's health was affected in some way by a lead pipe.

DID YOU KNOW?

Leachable lead is the amount of lead found in ceramic ware that was traditionally coated with lead glaze. The FDA didn't begin enforcing a set limit on it until about fifty years ago. This doesn't mean old Fiestaware passed down for generations necessarily contains lead, but it's still a good idea to purchase lead testing kits for any old pieces you might still be eating from.

TL;DR VERSION

- Pipes and plumbing have been around for more than five thousand years.
- While the dangers of lead are widely known today, it was a popular material used for water systems in the Roman Empire.
- While the character of Mr. Boddy in the movie *Clue* didn't die from the lead pipe, his health may have been affected by drinking water from lead pipes, eating from dishes and playing with toys decorated with lead glaze and paint, and driving a vehicle using leaded gasoline.

THE PAPER BAG

If you think paper bags are boring, you're probably right. No one really gives much thought to them, unless they break open while carrying groceries. But their history is much more complex than the bag itself.

Up until the 1800s, people would carry their items in fabric sacks, paper tied with twine, or straw baskets or cornucopias. When Francis Wolle was working at his father's grocery store in Jacobsville, Pennsylvania, in 1852, he began working on a machine that could quickly fold paper into an envelope-like bag. He formed the Union Paper Bag Company in 1869 and the use of his machine quickly became the go-to method for paper bag manufacturing.

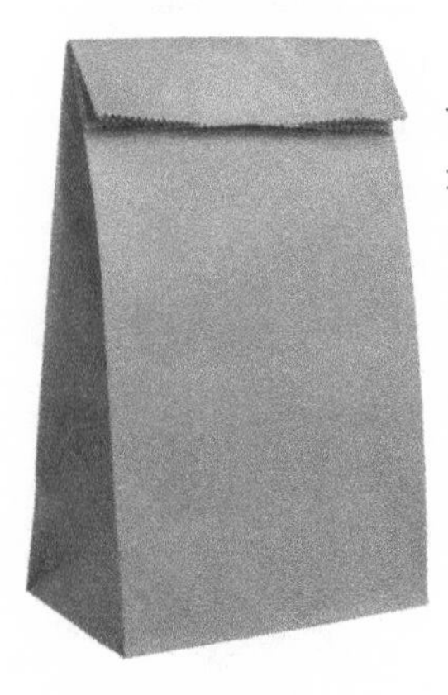

Margaret "Mattie" Knight was working for another paper bag manufacturing company, Columbia Paper Bag Company in Springfield, Massachusetts, making the envelope-style bags, when she began to hear rumors of attempts by engineers to create a machine capable of making a bag with a flat bottom.

As a child, she was known for tinkering with toys like kites and sleds to improve them, and for her inventive and inquisitive nature. She left school at twelve to begin working in a cotton mill in New Hampshire. When she saw the number of injuries that occurred at the looms when steel-tipped flying shuttles would come free, she devised a shuttle cover to prevent this. This cover ended up being used in not only that cotton mill, but across the Northeast. Again, she was only twelve when she came up with this!

When she heard the troubles other engineers were having inventing a machine to create flat-bottomed paper bags, she took on the challenge. While still doing her job during the day, she would study the machines, then leave work to spend long nights conceptualizing her idea for a machine that would use what she called a "plate knife holder," which would fold the bottom together. She finally completed a wooden model in 1867 to show her superiors.

During its demonstration, one thousand bags were folded—basically doing the work of thirty men—and she was given permission to take her machine to Boston to have an iron version created for use.

Here's the story takes on a movie-like plot twist, though. A Boston inventor by the name of Charles F. Annan was a frequent visitor to the machinist who was working on Knight's design. When Knight's machine was finished, she sought a patent for it in 1871, only to find out Annan had copied her design and had already filed a patent.

In a time when patent disputes weren't generally settled on behalf of females, Knight took Annan to court, spending $100 per day on a patent attorney to help defend her. Some sources even say that Annan's defense was that Knight wasn't properly educated and she couldn't possibly have come up with the concept of such a machine. With the help of all of her sketches, testimony, and eyewitness accounts, Knight was able to win the suit and was granted the patent. (Seriously, please, someone make this a movie!)

Even though Queen never sang about flat-bottomed bags making the rockin' world go round, Queen Victoria did decorate Knight with a Royal Legion of Honour that year for her innovation. By the time she died at the age of seventy-six, she had accumulated at least twenty-seven patents, including a machine for cutting shoe soles, a fire extinguisher, and various engine improvements.

But the bag didn't stop there. A few years later, Charles Stilwell improved upon the design by adding pleated sides, making it easier to lay empty bags flat for better storage. He named it the "SOS," which stood for Self-Opening Sack.

Stilwell's design is the basis for the modern-day paper bag, which fell out of favor when plastic bags became easier and cheaper to produce. Today, both bags are becoming more unpopular as some stores charge for disposable bags, prompting customers to bring their own instead.

However, depending on the studies you read, it's a toss-up between paper and plastic as to which is better for the earth. Plastic bags use fewer resources to produce, while paper bags disintegrate faster. More people are likely to recycle paper bags or reuse them (book covers, anyone?), while plastic bags are recycled only 1 percent of the time in the United States (higher in Canada), even though many plastic-bag manufacturers are now making their bags more environmentally friendly.

DID YOU KNOW?

Because Knight became so well-known for her inventions in her lifetime, she was heralded in her obituary as "Lady Edison." However, she died with only about $300 to her name.

TL;DR VERSION

- Before the flat-bottomed paper bag was invented, groceries were carried home in fabric sacks, wrapped packages, or large paper envelopes or cones.
- Margaret Knight created the flat-bottom bag machine, only to have the design stolen before she could take a patent out on it.
- She finally won the lawsuit over the patent and went on to live the rest of her life as an inventor—and took out patents as soon as she could on subsequent inventions.

SOLAR PANELS

As long as there has been life on earth, its reliance on the sun has persisted. Ancient religions worshiped the sun, including the Egyptians who worshiped Amun, the god of the sun, as the most supreme. The rulers of that time incorporated the name of the god into their own names because he was so revered (Tutankh*amun*, being the most noteworthy example).

The sun's rays have been harnessed for human use as far back as the seventh century BC, when a magnifying glass was used to create fire. Since then, humans have looked to the sun to provide not only light but also warmth, and eventually electric energy.

French scientist Edmond Becquerel discovered that the light produced by the sun created electrical voltage in 1839, when he placed gold or platinum in a solution and found an electrical current could be generated when those metal plates were exposed to the sun. This started a race in the scientific community to create a system in which solar power could be converted into electricity for everyday use.

However, the cost of materials made early attempts at commercializing solar power difficult. As early as the 1860s, August Mouchet developed a solar-powered printing press, but the French government terminated his funding, deciding it wasn't economically feasible.

Around the same time, electrician Willoughby Smith was using bars of selenium to help detect flaws in a cable that was used for telegraph communication. The selenium worked at night, but when the sun came out, it was barely effective. Realizing the sun affected its performance, Smith experimented to see if it was the light or the heat that was the key to selenium's conductivity. He, as well as other scientists, discovered that the material was creating a flow of electricity.

Once fossil-fueled power plants became the norm, though, solar energy research was put on a bit of a back burner. But George M. Minchin decided it was unsound for science to dismiss the sun as a source of energy and began to run a number of experiments during the 1800s.

Albert Einstein expanded Minchin's research and discovered that light consisted of packets of energy, which are now called photons, which varied depending on the wavelength. The problem was being able to convert enough photons efficiently enough to create energy for daily use.

But scientists didn't give up hope of using the sun as a source of power. One of the most noteworthy inventions was made by Maria Telkes, who worked at Massachusetts Institute of Technology during World War II, to create a solar-powered distiller that could turn seawater into drinking water.

She continued to work with solar energy throughout the rest of her career. In 1948, with the help of architect Eleanor Raymond, she built the first house heated with solar energy, which was located in Dover, Massachusetts. Not only was she able to capture solar rays to heat the home through boxlike panels, but she was able to improve upon past technology to develop solar-powered stoves and heaters. She also helped the Department of Energy build the first solar-electric residence in 1980.

As science in this area progresses, the capability of solar panels and solar power increases, and many are now supplementing their electricity consumption with solar power, with solar panels popping up on top of buildings and in large power-plant arrays around the world, trying to phase out fossil-fuel consumption, or at least reduce an expensive home electricity bill.

And if this quick chapter where this B-at-best science student tries to talk about technology doesn't intrigue you enough, there's always the cult classic so-terrible-it's-amazing movie *Birdemic* that may have convinced at least three or four people to switch to solar panels (or "slr pnls," as they're pronounced in the movie).

DID YOU KNOW?

If you're a podcast fan and want to find out more about solar panels and solar power, I highly recommend *Sunshine & PowerCuts* by Heather Welch, who lives entirely off the grid in New Zealand.

TL;DR VERSION

- Egyptians used to celebrate the sun god Amun and leaders would take on the name as part of their own, as in Tutankhamun.
- Albert Einstein discovered that light consisted of packets of energy, now known as photons.
- Maria Telkes built the first residence heated entirely with solar power.

TISSUES

Before there were tissues, there were handkerchiefs, which were used as far back as the ancient Egyptians in 2000 BC who carried white linen. The first written reference to the handkerchief was made by the Roman poet Catullus in the first century AD. They were most commonly used to shield the face from the sun and to wipe off sweat.

By the thirteenth century, they were known as a "couvrechef" in Old French—couvre meaning "cover" and chef meaning "head." This was soon renamed handkerchief by the English, as in Hand-Cover-Chief.

When Catherine de Medici married Henry II, she brought to France the fashion of lace-bordered and scented handkerchiefs, making them the popular accessories to own, although Henry was said to have used them to clean his teeth.

Shakespeare even used a handkerchief as the catalyst in the tragedy of *Othello* when Desdemona's handkerchief was planted on Cassio by Iago, resulting in Desdemona's murder by Othello out of jealousy. (Sorry, should I have given a spoiler alert?)

One of the most storied users of the handkerchief was Marie Antoinette in the eighteenth century, who was said to have torn pieces of lace from her dress and undergarments in the journey from Austria to France to wipe her tears as she was being brought to marry Louis XVI. She brought handkerchiefs to the mainstream when she arrived, wishing to always have one at the ready for future tears.

But this is considered to be more folklore than fact. However, she did decree that handkerchiefs be as wide as they were long, leading to a standardization of size of ten by ten inches or 11.5 by 11.5 inches—sizes that are still common today.

Handkerchiefs, for a while, were considered taboo if seen outside of the clothing. They were usually stuffed in pockets, purses, up sleeves, or in bodices, but when the two-piece suit came into fashion by the nineteenth century, it became more customary for men to fold them up nicely and use them as pocket squares to add a pop of color to an otherwise drab suit.

A woman would give a scented handkerchief to a man she showed favor to or, as the common portrayal went, she would intentionally drop her handkerchief for a knight or potential suitor to retrieve. By the way, they weren't always scented as a way for the owner to smell good; they were also used as face masks before modern-day plumbing was available.

Handkerchiefs became even more romantic during World War II, when European and American soldiers would keep them as tokens of affection, but they were just as useful when it was common for maps to be printed on silk handkerchiefs for pilots during both world wars in the event they were shot down.

As you can imagine, using a cloth to wipe one's nose resulted in easily spreading germs, not to mention having to carry around a snotty handkerchief in one's pocket or bodice.

When Kimberly-Clark Corporation came out with Kleenex in 1924, they were originally intended for use with cold cream to remove makeup. It wasn't until two years later that customers wrote to the company, telling them they used these facial tissues for blowing their nose more than any other use.

When a test was conducted in a Peoria, Illinois, newspaper asking readers to pick which use they favored, readers responded with 60 percent saying they used them as a disposable handkerchief for blowing their noses, leading Kimberly-Clark to change their marketing strategy to reflect that.

During World War II, when rations limited the manufacturing of Kleenex, the technology was used to help create bandages and wound dressings. By the time the war had ended and production for tissues was back in place,

Kleenex had become a household name, but not every tissue is a Kleenex. This is one of those brand names that became generic over the course of time.

The name Kleenex refers to its use, to "clean" the face, and the "ex" was a tie-in to Kimberly-Clark's other popular product, Kotex sanitary napkins. By the way, Kotex was a portmanteau of the words "cotton" and "texture," to describe the materials used.

DID YOU KNOW?

There is an internet urban legend about a supposed haunted Kleenex commercial from Japan that features a baby dressed as an ogre, and after there were some complaints about the commercial, a rumor started that the cast and crew that worked on it had either died or had tragedy strike them. None of that is true, but you can still find plenty of versions of this cursed Kleenex commercial on YouTube, including ones that apparently become distorted if watched at midnight.

TL;DR VERSION

- Handkerchiefs have been used as far back as 2000 BC.

- Catherine de Medici brought handkerchiefs to France when she married Henry II, but it was Marie Antoinette who popularized them.
- Disposable tissues were originally made to wipe cold cream off faces, but consumers used them more as a more sanitary way of blowing their noses.

THE TOOTHBRUSH

In a satire by Horace Miner from 1956 called "Body Ritual Among the Nacirema," the anthropologist wrote: "The daily body ritual performed by everyone includes a mouth-rite. Despite the fact that these people are so punctilious about care of the mouth, this rite involves a practice which strikes the uninitiated stranger as revolting. It was reported to me that the ritual consists of inserting a small bundle of hog hairs into the mouth, along with certain magical powders, and then moving the bundle in a highly formalized series of gestures."

Nacirema, by the way, is American spelled backward. Miner wrote about Americans and their so-called bizarre rituals, like brushing their teeth and going to the dentist, from an anthropological perspective. But if you weren't aware that this is a satire, it might sound completely bizarre.

But it's something we do every day, hopefully twice a day, as prescribed by dentists. But, weirdly, we're the only species that actively brushes its teeth.

You might have heard an anecdote or seen pictures about crocodiles who fall asleep with their mouths open while Egyptian plover birds pick at their teeth for remnants of their meals, essentially performing dental hygiene for them. According to ornithologists, no evidence of this actually exists, aside from CGI video from a chewing gum commercial, but animals who consume a lot of fiber naturally clean their teeth. Elephants even clean their tusks when they use them to dig hole or gnaw at tree bark.

So, why do we humans need to worry so much about our teeth?

Simple: Sugar and processed food deteriorate our teeth faster than the protein, grains, and fat eaten by many other animals. We're the only ones who can truly enjoy a Starbucks Unicorn Frappuccino, much to the detriment of our teeth.

Toothbrushes, as we know them today, came about fairly recently, compared to how long people have roamed the earth. It's believed humans as early as Egyptians in 5000 BC used their index fingers to brush their teeth, using a tooth powder made up of ash from ox hooves, myrrh, eggshells, and pumice.

By 3000 BC, humans were chewing on sticks with frayed ends, aptly known as chew sticks. Egyptians were so concerned with the brightness of their smiles, it was even common for them to use urine as a natural whitener.

Writings from thirteenth-century Chinese monks suggest they used a brush made of horse hair to clean their teeth. This wasn't the only brush of animal hair used. In the

Middle Ages, the west used horse hair, while boar bristles were used in the East. Goose feathers, silver, and copper were fashioned into toothpicks and used to clean teeth, as well.

But the actual word "toothbrush" wasn't recorded until 1690, in an autobiography written by Anthony Wood, an English antiquary—someone who studies history through artifacts. The toothbrush wasn't an artifact he found; instead, he mentioned it in his autobiography, in a passage about having to purchase one.

The first incarnation of what we know now as a toothbrush came when a rag maker named William Addis was sent to prison in 1780 for starting a riot. While there, he tired of using a rag and some brick dust to clean his teeth. It's said he was inspired by a broom to take an animal bone from his meal, drill holes in it, and have a guard procure him some coarse animal hair, which he either glued or wired through the holes.

Historians debate whether he really made the toothbrush in prison or just had the idea there, though. Either way, once he was released, he made a business of selling these toothbrushes, and the company is still around today, now known as Wisdom Toothbrushes.

By the Industrial Revolution, toothbrushes were a sign of status. Only the rich spent money on frivolities like keeping their teeth clean, and even then, there were no dental recommendations for brushing frequency, so there's no real way to know how often teeth were brushed back then.

It wasn't until school became mandatory for children in the early 1900s that tooth decay and, more importantly, its prevention, was brought to the collective conscious.

In a study of schoolchildren in Elmira, New York, only twenty-two out of 447 children had perfect teeth, while the rest had an abundance of cavities and decay. Dentists were brought into schools to show children proper dental hygiene. Schools began giving away toothbrushes or selling them cheaply, hoping the habit would catch on. That, combined with a successful advertising campaign by Pepsodent, defining a problem people didn't even know they had, which was a pesky, unattractive film on the teeth, with the easy solution to a beautiful smile, brought brushing into the mainstream.

When DuPont Laboratories invented nylon in 1937, bristles made with the new material replaced the boar or horse hair normally used for toothbrushes, and this was the same year the electric toothbrush was invented, which pretty much brings us to today.

So, here's a question, how badly do you want to brush your teeth now?

DID YOU KNOW?

In 2003, the toothbrush was designated the No. 1 invention Americans couldn't live without.

TL;DR VERSION

- In ancient times, teeth were cleaned either with one's finger or by chewing on twigs.
- Humans are the only species that need to brush their teeth because of the sugar and processed foods we eat.
- Ragmaker William Addis came up with the idea of the toothbrush while in prison in 1780.

PART SIX:
MUSIC

LULLABIES

One of my podcast mentors, Dave Jackson, used to say that babies were podcast killers. When I started *The Story Behind*, I was determined not to let my pregnancy deter any plans of continuing the show, and I was able to store up enough episodes to last five weeks. But, when I came back, a lot of topics I wanted to cover revolved around my new world of everything-baby, including this one, which I found out was one of the ways my son could be calmed.

Lullabies were around as early as 2000 BC, as far as we know. On a palm-sized clay tablet dating back to ancient Babylon (modern-day Iraq) is a lullaby written in cuneiform script. Although meant to soothe a baby to sleep, the song's lyrics warn the child to go to sleep or else a demon will wake up and eat him.

If you think that's the most menacing thing one could sing to their own child, know that similar warnings are sung to children, even in modern times, all over the world.

Take a look at the tragedy of a tree branch breaking and a baby falling from it that we mindlessly sing in "Rock-a-bye Baby." Since many lullabies are passed down from generations ago, it's easy to forget that songs, nursery rhymes, and even fairy tales weren't always as family-friendly as they are today, with some being downright morbid and scary. Of course, since babies don't understand language yet, it's the soothing sounds of these songs being sung that help soothe the baby.

When a baby is in the womb, there is little from the outside he can sense. In fact, the reason noise machines for infants featuring white noise and heartbeats are so popular is that they are meant to mimic what the fetus experiences as its hearing develops in the womb.

But the other sound a baby hears before birth is the mother's voice. Even though it's muffled, they are still aware of it, and it's said lullabies sung by the mother can act as a bridge between life in the womb and life outside of it. This can also work if, while in utero, the baby is exposed to others' voices on a consistent basis. (From the number of podcasts I listen to, I often wondered which podcast hosts my son is most familiar with, actually.)

There's even a story of a brother who sang "You Are My Sunshine" every night to his baby sister while she was in his mother's tummy, and when the baby was born with health problems and didn't look like she would make it, it was her brother singing to her that brought her vitals up and caused her to make a miraculous recovery. Snopes has since given this Legend status, though.

One thing I didn't do with my children was to sing the most basic lullabies—the ones everyone knows—mostly because I'd heard them so much that I was sick of them before even having children. But when you think about these lullabies, do you ever wonder why they are effective?

Soothing tones and gentle rhythms seem to make the most sense for why a lullaby would help a baby fall asleep. And for those of you, like me, who can't stand traditional lullabies, CDs of more popular songs made into lullabies have become popular, like the *Rockabye Baby* CDs with lullabied covers of artists like Journey, Aerosmith, the Beatles, Kanye West and even the Hamilton soundtrack, making them more palatable to adults' ears.

However, studies have shown that having familiar voices sing these lullabies can be the most effective at soothing a child. Not only that, singing lyrics to songs or even reciting nursery rhymes can help support early language skills. Music itself has been proven a beneficial aid to learning for babies and young children, as well. In contrast, while you may have heard of the "Mozart Effect"—the idea that playing classical music for babies can make them smarter—scientists have had mixed findings on this, and even warn against using classical music as a substitute for general play and interaction with parents.

The lyrics for the Beatles' "Good Night" or "Golden Slumbers" may make more sense as a lullaby, but the songs "Lucy in the Sky with Diamonds" or "Norwegian Wood" might be more effective because of their 3/4 time signature. Commonly known as the waltz time signature, it mimics rocking, and many traditional lullabies from all over the world are written in it. The rocking motion and even beat of the lullaby mimic the motion the baby felt before being born.

One of the most familiar lullabies is Brahms' "Lullaby," also known as the "Cradle Song." It was written in 1868 and originally called *Wiegenlied: Guten Abend, Gute Nacht,* which is German for "Good Evening, Good Night," and one theory was that it was based on a Viennese song a woman named Bertha Faber used to sing to Brahms.

It's unclear if he had a romantic relationship with her when they first met, but when he rekindled his acquaintance with her after a number of years, after she was married and had a child, he sent the song to her with the intention that while she was singing her son to sleep, she might realize he had written it as a love song for her.

The song was also recorded by a representative of Thomas Edison in 1889 and is considered the earliest recording made by a major composer, even though the piano is mostly inaudible because of heavy surface noise.

As for "Twinkle, Twinkle, Little Star," it's commonly attributed to Mozart, but the melody actually comes from an old French song. Mozart did compose a set of twelve variations based on the melody, most likely while he was in Paris in 1778, and it was given lyrics from a poem called

"The Star" by English poet Jane Taylor. It's unclear if Taylor intended her poem to become the lyrics to the song when she originally wrote it, but it appeared together with the melody in a songbook published years later.

The origins of "Rock-a-bye Baby" and "Hush Little Baby" are both varied and hard to pinpoint, but the R&B version of the latter, written and recorded by Inez and Charlie Foxx and popularized by James Taylor and Carly Simon, became more of a novelty song, rather than an actual lullaby. It's difficult to even think of that song now without picturing the rendition sung by Jim Carrey and Jeff Daniels in the 1994 movie *Dumb & Dumber.*

DID YOU KNOW?

And, in case you want some ideas for lullabies, one of my favorite Spotify Playlists I've put together for my son can be found at thestorybehindbook.com/LullabyPlaylist.

TL;DR VERSION

- Lullabies were around as early as 2000 BC.
- One way to distinguish a lullaby is by its time signature. Many songs we consider lullabies are written in 3/4 time.
- Brahms' "Lullaby" was actually written as a love song to a married woman to sing to her child.

MUSICALS

This is probably my favorite episode of The Story Behind,
since it was entirely was sung to the tune of
"The Major-General's Song" from The Pirates of Penzance.

This is the story all about the modern major musical.

The history's quite large and certain facts might
be disputable.

It started back in Greece where many theaters
were acoustical.

The Middle Ages brought them back with church and
tra-ve-ling minstrels.

The 17- and 18-hundreds' shows were quite satirical.

Though opera was
serious, the musicals
were comical.

The views expressed
may not have always
suited those imperial.

But entertainment won and they continued to be fashionable.

With names like Strauss and Offenbach, they spread
fast geographical.

They hit the USA and folks enjoyed a bit of Vaudeville.

The history's quite large and certain facts might be disputable.

This is the story all about the modern major musical.

Burlesque, we will not cover because this podcast is for all ears.

But just to note, this was a part of musicals for several years.

Extravaganzas, pantomimes, and farces were all met with cheers.

The theater thrived and fattened wallets of investing financiers.

Broadway emerged as the grand force to rule all things theatrical.

And stage revues like Ziegfeld's follies drew crowds international.

In 1919, unions forced their contracts with a picket line.

The '20s saw Cole Porter, Hart and Rodgers, and then Hammerstein.

The Great Depression didn't keep the musicals from going down.

But actors moved out west to Hollywood for movies now with sound.

The history's quite large and certain facts might be disputable.

This is the story all about the modern major musical.

The people went to movies to escape thoughts of the second war.

The musicals were loved and records flew off shelves at music stores.

The decades passed and rock 'n' roll soon stepped on music theater's toes.

But Sondheim came and introduced the people to his concept shows.

The trend revived and musicals came back strong like they were before.

The audiences clamored for Andrew Lloyd Webber's epic scores.

And now our favorite movies-turned-to-musicals all do quite well.

And history has sold out crowds with R&B by Lin-Manuel.

For Episode 100 I came up with doing musicals.

I hope you learned a bit and all my rhymes weren't inexcusable.

The history's quite large and certain facts might be disputable.

This is the story all about the modern major musical!

DID YOU KNOW?

White electric bulbs were used to illuminate Broadway in the 1890s, giving it the nickname "The Great White Way."

TL;DR VERSION

- Musicals were around since ancient Greece.
- Theater and music were often used as satire to mock those in power.
- Musicals go through waves of popularity, most recently with *Hamilton* becoming a hit.

THE THEREMIN

If you haven't heard of a theremin, you've certainly heard the sound it makes. The most common use of it is in the music and sound effects from sci-fi or horror movies. It's a little more difficult to describe it by writing, as opposed to playing the sound on *The Story Behind* podcast, but it basically sounds like the musical equivalent to this: "Wuaaaaaaahhhh!" (Helpful, no?)

The theremin may be one of the weirdest-sounding instruments, and to watch someone play it is like watching a wizard performing an intricate spell or a conductor in front of an invisible orchestra. This may sound like a more recent invention, but it's been around for about 100 years, believe it or not.

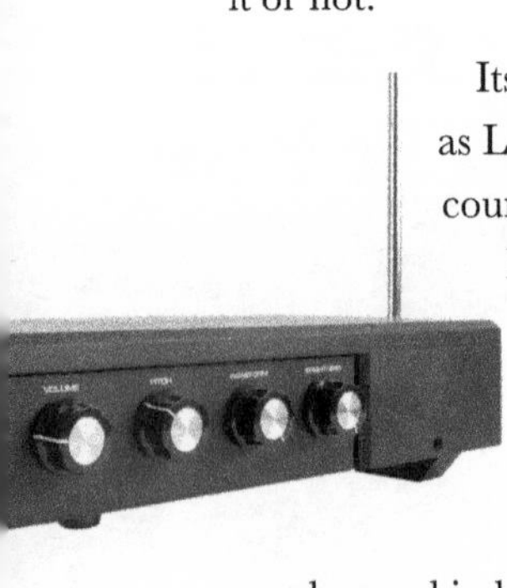

Its inventor, Lev Sergeyevich Termen (known as Leon Theremin in the English-speaking countries), was born in Russia in 1896. During the First World War, he was in college studying physics, and played the cello in his free time. He progressed quickly through his studies.

When his time to serve had come, he was so advanced in his studies he was able to avoid the front lines and work as a scientist and engineer.

He began using electromagnetic fields to study the density of gases during the Russian Civil War that began in 1917 and lasted until 1922. The machine he created, a dielectric device that worked as a voltmeter, would make sounds at different frequencies as he moved around near it. Being trained as a musician for a number of years prior, he realized he could make music by moving his hands ever so slightly between two electromagnetic fields around the two antennae of the machine.

With some adjustments and practice, using one hand for pitch and the other for volume, he perfected playing what became known as the theremin in 1920—the first instrument that could be played without touching it.

When he showed his design to Vladimir Lenin, Lenin was so impressed he sent Theremin and his theremin on a tour of the country to show off the instrument. Lenin wanted to spread the news of what could be done with electricity, after all one of his favorite maxims was "Socialism equals Social Power Plus Electrification."

Theremin was sent over to the United States in 1928 to debut his new instrument and to hopefully sell it to a company to mass-produce.

When Theremin got to the US, RCA was interested in the instrument and had him sign a contract for $100,000. The theremin debuted, but initially failed to sell. One of the main reasons for this was the debut came just 10 days before the stock-market crash of 1929.

One day in 1938, Theremin vanished, leaving his wife and his business failures behind him in the United States, only to reappear years later in Russia after the fall of the Iron Curtain. Stories vary, as far as the reason he left America—some say he was homesick or wanted to escape debt; others say the NKVD (the predecessor to the KGB) was responsible for kidnapping him. Back in Russia, he was arrested and imprisoned for reasons that are still unknown.

He admitted at the age of ninety-five that debuting the instrument had been an excuse to come to America as part of the development team working on a Soviet surveillance device. He also admitted that once he installed the device in a government building, all he heard were boring conversations and no government secrets.

The theremin was all but forgotten until Hollywood began using it in the soundtracks to science fiction movies in the 1950s and '60s, such as *The Day the Earth Stood Still* and Alfred Hitchcock's *Spellbound.*

After that, there was a resurgence in the popularity of the instrument, and a man named Robert Moog began selling build-it-yourself theremin kits. Moog is now known as the man behind the Moog synthesizer, but the theremin is still one of the company's best-selling items.

You may think that sliding instrument you hear in the Beach Boys' "Good Vibrations" is a Theremin, but it's actually an electro-theremin, an easier-to-play alternative created by Bob Whitsell and played by Paul Tanner.

In Led Zeppelin's "Whole Lotta Love," Jimmy Page uses a theremin-like instrument during the solo, as well as throughout "No Quarter."

The theremin nowadays is still considered one of the most unusual instruments. In an episode of *The Big Bang Theory*, Sheldon Cooper is seen playing a theremin, which he says he has enjoyed since hearing the original Star Trek theme. That's a common misconception, though (yes, believe it or not, sometimes *The Big Bang Theory* gets things wrong). That theremin sound is actually singing by renowned soprano Loulie Jean Norman.

DID YOU KNOW?

While the stock-market crash was definitely a factor in the failure of the theremin when it first debuted, RCA also made the advertising faux pas of suggesting it was easy to play. The few buyers of the instrument soon found out waving one's hand over the electromagnetic box to produce music literally out of thin air was more difficult than they were promised.

TL;DR VERSION

- Leon Theremin was studying electromagnetic fields in 1920 in Russia when he came up with his instrument, the theremin.

- Lenin sent Theremin to the United States under the guise of promoting his invention, when in reality, he was actually sent on a mission to hide surveillance devices in government buildings.
- Sometimes, *The Big Bang Theory* gets nerdy facts wrong.

PART SEVEN:
OUT & ABOUT

THE FIRE HYDRANT

As long as fire has been around, so has the need for a way to put it out. For years, firefighters used bucket brigades, moving buckets of water from a nearby lake or pond to the source of the fire. During the 1600s in London, hollowed-out logs were used as pipes and firefighters would dig to get to the water source when needed, plugging up when the water was no longer needed. These wooden plugs are still referred to today when people call fire hydrants "fire plugs."

When a fire destroyed three-quarters of London in 1666, the city began putting pre-drilled holes and plugs in their water system, which they then kept above ground. Soon cast-iron pipes began replacing the wooden ones in water systems in larger cities.

Most historical references list Frederick Graff Sr. as the inventor of the above-ground fire hydrant—the one we're most familiar with today—although the patent for his invention can't be verified because, in an ironic twist, the US patent office burned to the ground in 1836.

The carpenter-turned-engineer was instrumental in creating Philadelphia's first water system, which included the pillar-shaped hydrants for easy water access for firefighters. Although his initial system was found to be inadequate as the years went on, he kept trying to perfect it, finally succeeding in 1822.

While we recognize fire hydrants as being made of metal today, the first ones were metal pipes encased in wood with a valve at the bottom and an outlet on the side. A common problem for early hydrants was keeping the water inside from freezing during the colder months. Workers tried to insulate the pipes with sawdust or manure, but it didn't prove to be very effective.

An adjustment of the valve location inside the hydrant to lower it below ground level was finally made to allow the water above the valve to empty at the end of each use. The nut to turn the valve on is usually seen on top of the hydrant. This is known as a dry-barrel hydrant, while in areas where freezing isn't likely, water is flowing to a hydrant at all times and these are known as wet-barrel hydrants.

If you've ever wondered why some hydrants are red and some are yellow, this isn't an aesthetic choice, it's actually a way to differentiate the pressure of the water that flows through the hydrant. The pressure ranges from 500 gallons per minute to more than 1500 GPM. The more densely populated the area, the more pressure is needed, since a fire would more likely spread faster and, in a busy city, it might take firefighters longer to reach a fire.

Colors vary by city and location; however, a common color system is that yellow hydrants are connected to public water supply of cities and red is used for more rural areas or special operations. If you see a purple fire hydrant, it most likely means the water comes from a lake or pond nearby and is considered non-potable. Blue and green are usually used in metropolitan areas and the water pressure is on the higher end of the spectrum.

If the flow of a hydrant isn't high enough, the water system can suck in water from the ground, which may be polluted and untreated, leading many to get a Boil Water Advisory.

DID YOU KNOW?

Fireman George Smith installed the first fire hydrant in New York City in 1817 in a pretty convenient location for him: right in front of his house.

TL;DR VERSION

- Before fire hydrants, bucket brigades were used by firefighters to supply water to put out fires.
- The patent for the first fire hydrant cannot be verified because the US patent office burned down in 1838.
- The different colors you see on fire hydrants refer to the water pressure, measured in gallons per minute.

GAS MASKS

The gas mask is probably best known for its use during World War I to fend off mustard gas attacks. But gas masks have been around much longer, although usually they were used to protect the wearer from obnoxious smells and fumes. One of the earliest known masks to shield the nose and mouth from fumes was the simple sea sponge, which was used in ancient Greece.

In *The Book of Ingenious Devices,* written in 850 AD, three brothers living in Iraq, known as the Banu Musa, listed their invention of a mask workers used in polluted wells.

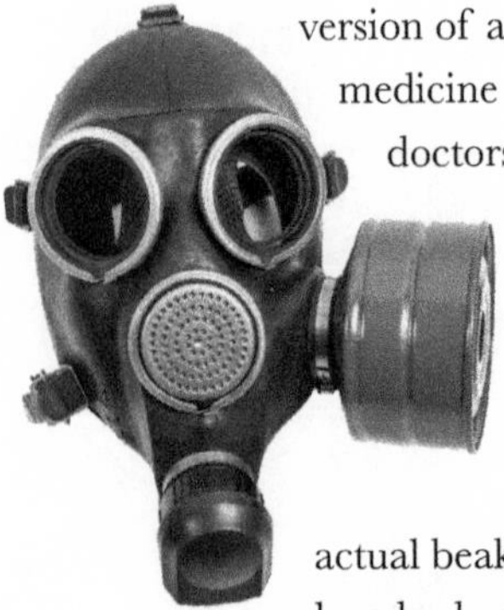

The beaked mask is considered another early version of a gas mask. Before modern medicine and the discovery of germs, doctors believed illness was caused by noxious bad air, which was known as miasma.

The mask is now associated with what's known as the Plague Doctor ensemble; however, the actual beaked mask was created three hundred years after the Black Plague.

Doctor de Lorme, who was the chief physician to Louis XIII, designed the mask in the seventeenth century. It was also attached to what is considered an early hazmat suit, since the accompanying outfit's leather overcoat was tucked underneath the mask, and its breeches, gloves and boots left very little exposure to outside air.

The reason for the beak was to stuff dried flowers and herbs inside for the doctor as a way to breathe more pleasant-smelling air as he worked on patients in a time before bathing was more common.

Another mask began following a parallel timeline as the gas mask by the early 1800s: the diving mask. While gas masks were created as a way to purify the air for the wearer, the diving mask was used to go underwater and oxygen was pumped into the breathing apparatus via a tube. But some gas masks borrowed from this design, especially for situations in which the air was contaminated and a respirator was needed.

The first US patent for an air-purifying respirator was given to Lewis P. Haslett in 1849 for a device which was able to filter dust from the air. A few years later, Scottish chemist John Stenhouse used a filter made of charcoal, which is still common today as a way to filter air, as well as water (like in your water filtration pitchers, for example).

One noteworthy mask was developed by Garrett Morgan, who is also mentioned in our "The Story Behind the Traffic Light" chapter.

Morgan made national news when he used his gas mask and safety hood to rescue thirty-two men trapped by an explosion in an underground tunnel in 1914. In the same year, mustard gas and a lung irritant, dianisidine chlorosulfate, were used by the Germans in World War I, nullifying the international treaties signed years before against using poison and poisonous weapons in war.

Morgan's design is credited as the basis for the army gas masks used during World War I, which included a hood to also protect the face and eyes. Before that, soldiers used urine-soaked rags over their faces.

The following year, hundreds of Allied soldiers died after Germans used chlorine gas during the Battle of Ypres. A Scottish medical researcher named John Scott Haldane was tasked with finding out exactly what kind of gas was used. He did so by examining the effect the gas had on the metal buttons worn by Allied soldiers.

Because his lab was a part of his home, he wanted to make sure his family was safe from his experiments. He enlisted his teenage daughter Naomi to stand outside the door to his lab and watch through the window. Should any of the researchers become incapacitated, Naomi was to run in and grab them as quickly as she could in order to resuscitate them.

When Haldane was able to identify the chlorine gas, he began making a respirator that used cotton and gauze soaked in a solution that neutralized low concentrations of the gas. His contribution, which later added a respirator tank, was the beginning of modern respirators.

But, if you want to look at sci-fi history, we all know the respirator was most infamously used a long time ago, in a galaxy far, far away, by a certain Lord of the Sith.

For those looking to put together a nuclear attack survival kit, two types of gas masks are recommended. One known as NBC-approved, which stands for protection against nuclear, biological, and chemical gases, and the other is CBRN-approved, which is the same as NBC, but also includes radiation.

One key element in selecting a gas mask is to make sure the fit is snug and creates an airtight seal around the face. Beards are considered a hindrance to gas masks because they can disrupt the seal. Some commercially available masks even come with a device to drink through.

DID YOU KNOW?

Gas masks are not designed for long-term wear. The filter needs to be changed every three to four hours.

TL;DR VERSION

- The first known mask to protect the wearer from noxious fumes was a simple sponge used in ancient Greece.
- The beaked mask, commonly known as part of the Plague Doctor ensemble, was actually created three hundred years after the Black Death plague.
- When shopping for a gas mask, make sure it fits snugly around your face, and beware of overgrown facial hair that could get the way of a tight seal.

MAD HATTERS

Most people know mercury as a toxic substance today, but throughout the twentieth century, the chemical compound mercury nitrate was used in the hat-making industry. Mercury nitrate made the outer hairs of a pelt soft and limp so they could be pressed together easily to make felt. This process was kept a secret by the French until the Huguenots were forced to flee from France in the 1600s. When they came to England, they shared the secret and it became widespread.

This became known as carroting, since the process turned pelts of white rabbit fur to a reddish orange color.

Mercury was used to treat a number of medical problems, including syphilis. Physicians were seeing patients who were treating themselves with a mercury ointment exhibiting similar symptoms of excessive salivation and gingivitis, and the psychiatric symptoms of shyness, loss of self-confidence, anxiety, fear of ridicule, and an explosive loss of temper. The syphilis and treatment were

described in the eighteenth century as "A night with Venus" being "followed by the lifetime with Mercury."

By the eighteenth century, due to poor ventilation in hat-making factories and prolonged exposure to mercury nitrate, some workers began exhibiting similar symptoms: anxiety, depression, irritability, loss of coordination, slurred speech, and trembling, which became known as "hatters' shakes."

Danbury, Connecticut, was known the world over as Hat City because of the number of hats manufactured there, and even originated the term "Danbury Shakes" for the symptom. The Board of Health knew about the problem, but didn't do much to stop the use of mercury because it wasn't affecting anyone in the town other than hat workers.

The phrase "mad as a hatter" also became part of everyday speech to describe someone exhibiting similar symptoms to those of mercury poisoning.

One famous name believed to have suffered was Boston Corbett, a hat-industry worker responsible for leading his regiment of the New York Cavalry to track down John Wilkes Booth following the assassination of Abraham Lincoln. Corbett disobeyed orders to capture Booth and, instead, shot and killed him. Corbett was cleared of blame and declared a hero by many, but, years later, he was put into a mental asylum after pulling a gun and threatening a group of people at the Kansas Statehouse. He managed to escape the asylum and disappeared, never to be seen again.

It was in the 1940s that mercury use in hat making was finally banned in the United States.

It may seem to make perfect sense to use this common disease to explain the quirks of the character of the Mad Hatter in Lewis Carroll's *Alice's Adventures in Wonderland*, but the repetition of this theory led many to believe it was true when, in fact, the real inspiration for the character was neither mad from mercury poisoning, nor even a hatter.

Johnny Depp described his portrayal of the Mad Hatter in the 2010 film *Alice in Wonderland* as a victim of mercury poisoning, including the carroting effect you see in his hair and eyebrows, when he said, "I think he was poisoned—very, very poisoned. And I think it just took effect in all his nerves. It was coming out through his hair and through his fingernails, through his eyes."

But the movie portrayal and Johnny Depp's character analysis may have been a bit different from Lewis Carroll's original character. Remember when I said one of the traits of mercury poisoning was shyness and anxiety? The character of the Mad Hatter, both in the book and as described by Depp for the movie, was extremely extroverted. Carroll's Mad Hatter wasn't described as suffering from shaking, aside from constantly checking his watch and shaking it. This was more to show his obsession with time, which was explained in the book as a result of his singing being interrupted by the Queen of Hearts, exclaiming, "He's murdering the time! Off with his head!"

In fact, many scholars believe the Mad Hatter was based on an eccentric porter and cabinet maker named Theophilus Carter of Oxford, England. Carter was known for wearing a large top hat as he stood outside his shop, and locals began referring to him as "the Mad Hatter."

Carter became known for displaying the invention of an alarm-clock bed, which tipped the sleeper out of bed at the designated hour. This invention was displayed at the Great Exhibition in Hyde Park in 1851—actually, two different versions were displayed. Neither was credited to Carter, but the idea was somehow attributed to him over the years.

Carroll, whose real name was Charles Lutwidge Dodgson, would have been aware of Carter and his eccentricities. Carroll has denied any of his characters were based on real people, even though many believe Alice to be based on seven-year-old Alice Liddell, who once traveled on the same boat as Carroll. The boat was traveling to Oxford, where Carter happened to live and work, meaning Carroll would have met or at least heard of Carter.

When a photograph surfaced of Carter, scholars compared it to the illustrations from the original books, drawn by Sir John Tenniel, and noticed the same recessed chin and over-developed nose, adding to the belief that Carroll had based his character on Carter—but we may never truly know.

DID YOU KNOW?

The phrase "mad as a hatter" predates Carroll's book. One theory is that it came from an eccentric known as John Hatter in the seventeenth century. The other is another seventeenth-century hat maker by the name of Robert Crab from Chesham, England, who was the original "Mad Hatter," who was known for his overzealous religious delusions. Following a head wound suffered during the English Civil War, he proclaimed himself a prophet and donated his possessions to the poor to live on the docks.

TL;DR VERSION

- From the 1600s through the twentieth century, the majority of people didn't realize the dangers of mercury. It was even used for medical problems, including syphilis.
- It wasn't until the 1940s that hatmakers finally stopped using mercury in their process.
- We may never know for sure, but Lewis Carroll's Mad Hatter may not have been based on a hatter who was mad from mercury, but on an eccentric who was local to Carroll and was known for wearing a tall top hat.

THE SMILEY FACE

This chapter is based on an episode of The Story Behind *podcast that was part of a series based on pop-culture references in the movie* Forrest Gump.

In 1963, Harvey Ross Ball, a graphic designer who worked at the State Mutual Life Assurance Company, created the iconic smiley face when he was tasked to come up with a morale booster for the insurance company he worked for in Worcester, Massachusetts. It took him all of ten minutes to create the smiley face, which originally began as just a smile, but, as an afterthought, he added eyes so it couldn't be turned into a frown. He was given $45 for his work. Sounds like a dream, right? Forty-five bucks for only ten minutes of work? But that would be about all the money Ball would receive for the design, in the long run.

Soon, the smiley face began adorning posters, buttons and signs. But neither Ball nor State Mutual Life trademarked the design, and soon it attracted Bernard and Murray Spain, owners of two Hallmark stores. It was the Spain brothers who added the catchphrase, "Have a happy day," later changed to "Have a nice day," and copyrighted the design. They're credited with helping to boost America's mood following the Vietnam War with their jolly slogan.

America wasn't the only place where the smiley face was seen. In France in 1972, Franklin Loufrani began using the symbol in the publication *France Soir* to mark good news. He trademarked what he called the "Smiley" and began marketing transfers for T-shirts in more than 100 countries. The Smiley Company is run by his son, Nicolas, and is one of the top licensing companies in the world.

You can imagine, with the smiley face popping up all over the world at this point, the claim to be the originator has been fought over several times. The Spain brothers acknowledged Ball's original design; however, when credited as the inventors, they weren't quick to correct anyone.

Nicolas Loufrani has argued that the design is so simple, there could be no proof as to who the originator was. He has even referred to cave-art depictions of smiley faces dating back to 2500 BC on the company website.

But that didn't stop the Smiley Company from trying to trademark the symbol in 1997. Can you guess what company tried to stop them? Walmart. Walmart had been using the smiley face as part of the company's branding since 1996. The case was settled out of court, but not before both companies lost a lot of legal fees, not to mention ten years, embroiled in this dispute.

The unknown origin of the smiley face was probably what led writers of *Forrest Gump* to incorporate that little nod to the logo's popularity in the movie, but, a year before it came out, a Seattle man named David Stern ran for mayor, claiming in campaign ads that he was the inventor of the smiley face. The local newspaper, the *Seattle Post-Intelligencer*, was able to call his bluff, though. He lost the election regardless.

It seemed the smiley face was everywhere, and to Harvey Ball, it had lost all meaning. In 1999, Ball created the World Smile Corporation and World Smile Day, held on the first Friday in October to raise money for the Harvey Ball World Smile Foundation, which supports grassroots charities and various children's causes. Ball died at the age of seventy-nine in 2001, and the company he created the smiley face for, now called Worcester Mutual Fire Insurance, still uses his smiley face design on its promotional materials.

:)

DID YOU KNOW?

The smiley face can be seen in a *New York Herald Tribune* advertisement for the 1953 movie *Lili*, leading many to consider this its first appearance, even before Ball's design.

TL;DR VERSION

- Harvey Ball designed the smiley face for an insurance firm he worked for and was paid $45 for his work.
- Many tried to capitalize on the simple yet catchy design, including David Stern, who ran for mayor of Seattle with the claim he invented it.
- World Smile Day, created by Ball, is held on the first Friday in October.

THE TRAFFIC LIGHT

Ever feel like the world is conspiring against you when you're stuck at a red light? It may seem like an inconvenience, maybe adding a few extra minutes to your commute. But before traffic lights were as commonplace as they are today, the number of cars, wagons, bicycles, and pedestrians made getting around cities more dangerous and even slower.

But let's rewind.

When automobiles first appeared in the 1800s, it took a while for them to catch on. In fact, when the Ford Model T was introduced in 1908, only 20 percent of the roads in the US were paved. Cars didn't catch on as automobile makers had hoped, especially because many of the rules of the road we know today weren't established.

It was reported that in the first six months of 1906, more Americans were killed by drivers than by the Spanish in the Spanish-American War.

Some states passed laws in an effort to reduce the number of accidents, such as Vermont requiring every motorist to have a person walk in front of a moving vehicle with a red flag. Others took it upon themselves to show their hatred for motorists by creating log barriers on roads, scattering glass, or even adding ditches.

But driving did offer a faster, more efficient way for people to get around, and in 1908 when Henry Ford introduced the Model T and, with that, new mass-production techniques, cars became more commonplace. And, with that, so did automobile accidents.

As more horse-drawn carriages and buggies were being traded in for brand-new automobiles, the number of accidents increased, especially since many rules of the road were not established yet. There were very few traffic-guiding systems already in place around the world, like a signal outside the Houses of Parliament in London in 1868 with two arms that would alternately raise and lower, one reading *caution* and one reading *stop*, similar to a railroad crossing signal you might see today. At night, the signal used red and green gas lamps, but these lamps exploded a month after being installed, killing the policeman who was operating it.

In 1912, in Salt Lake City, Utah, a police officer named Lester Wire created a box connected to trolley and light wires for a signal featuring red and green lights. The first electric traffic signal was based on a design by James Hoge and was installed in Cleveland, Ohio, in 1914. It was operated from inside a control booth and is marked as the first electric traffic light.

Even in New York City, which is still infamous for its traffic congestion, it was said that to drive from Fifty-Seventh Street to Thirty-Fourth Street could take up to forty minutes in the early 1900s. This was annoying not only to motorists, but also for businesses. Luckily, in 1920, Dr. John A. Harriss designed a simple two-light signal, which was controlled manually by a police officer, and the twenty-three-block trip went from forty minutes down to nine.

The design caught on and, two years later, an architect named Joseph Freedlander was commissioned to design an ornate set of signals for intersections along Fifth Avenue, which featured a brass statue of Mercury on top, the Greek god of travelers.

But the inventor of the three-signal traffic light we're most familiar with today was named Garrett Morgan.

Garrett Morgan was born in 1877. He was one of eleven children born to Sydney and Elizabeth Morgan, who were former slaves in Kentucky. He had an elementary school education, but paid for tutoring once he began working. He moved to Ohio and took a job as a sewing-machine repairman. He was so successful, he opened his own repair business.

Before his contribution to the history of traffic lights, he patented a number of other inventions, including a chemical hair straightener he created as a result of his search for a solution for wool getting scorched when being put through sewing machines. He also patented a safety hood for breathing in smoke, which became a precursor to modern-day gas masks (also talked about in the "The Story Behind Gas Masks" chapter of the book).

Morgan was the first black man in Cleveland to own a car. At that time, Cleveland's streets were narrow, and traffic signals only switched between Stop and Go, leaving drivers and pedestrians little time to react to avoid a collision. When Morgan witnessed an accident between two carriages, he had the idea to create a third signal as a warning for pedestrians and motorists.

Thus, the three-signal traffic light was born. He received a patent in 1923 for his T-shaped design, which he then sold to General Electric.

If you ever wondered why red and green are used for Stop and Go, there are two theories.

One is that red and green lights on equipment were already common. Red indicated the machine was stopped and green indicated the machine was running.

The other theory is that red has the longest wavelength of any color on the visible spectrum, meaning it's easier to see from a distance, giving you more time to slow down to a stop.

Green, was originally used in railway lights to mean caution, but because the wavelength of the color is shorter than that of red and yellow (so it scatters more), it looks white from a distance, which caused engineers to mistake the green light for a white star in the distance until it was too late.

Yellow light is almost as easy to see as red, which is why the color is used in many caution signs like school zones or crosswalks.

DID YOU KNOW?

Originally, the driver's side was the right side of the automobile, as it was in many horse-drawn carriages. This helped drivers to avoid ditches along the side of the road. When Henry Ford began producing his cars, he moved the steering wheel to the left side for drivers to be able to see oncoming traffic better.

TL;DR VERSION

- When the Ford Model T was introduced in 1908, only 20 percent of the roads in the US were paved.
- In the early 1900s, it could take up to forty minutes to drive from Fifty-Seventh Street to Thirty-Fourth Street in New York City because of the lack of traffic signals.
- Garrett Morgan, the son of two former slaves, came up with the three-signal traffic light we're most familiar with today, as well as a chemical hair straightener and the predecessor of the gas mask, among other things.

WINDSHIELD WIPERS

As anyone with a car would know, without working windshield wipers, it can feel incredibly dangerous trying to drive in inclement weather. Believe it or not, though, one of the earliest patents for such a mechanism was granted to a woman who received rejection letter after rejection letter from car manufacturers for her idea.

Her name was Mary Anderson. She was visiting New York City from Alabama and riding in a streetcar during a sleet storm. Normally, when drivers were caught in bad weather, they would either open the window to remove any debris by hand or stop the vehicle and manually remove it. Both of these methods meant that either passengers got a face-full of cold air and unfavorable conditions in their faces, or it would lead to delays.

Anderson had the idea, as she watched her streetcar driver continually stop the vehicle and remove the sleet by hand, to create a mechanism that could be controlled from within the car to remove such debris. (Another version of the story has the driver opening his side window and sticking his head out to try to see better.)

That very same day, she began sketching a mechanism made of wood and rubber that drivers could use a lever to activate, removing the sleet and rain. She also designed her wiper to be removed for months when snow and hail weren't as much of a problem.

She patented the idea in 1903 and began trying to market it to early automobile manufacturers.

Nowadays, the windshield wiper is ubiquitous on nearly all automobiles on the road. However, Anderson's idea was met with laughs from peers and arguments that the sweeping motion of the arm would distract the driver. Automobile makers wrote back to Anderson, saying they couldn't find the commercial value of such a product.

A few years later, in 1916, a Buffalo, New York man by the name of John Oishei hit a cyclist while driving his automobile in the rain. The cyclist was uninjured, but the accident shook up Oishei so much, he knew there had to be a safer way for motorists to see in the rain.

By that time, other forms of rubber wipers were on the market, although not popularized yet. When Oishei came across a hand-operated one known as the Rain Rubber, he founded a company to market it.

It was a bit cumbersome and required a driver to be able to manually remove the debris with it with one hand, while the other hand was still in charge of steering and shifting the car, but it still found its market.

The story, however, doesn't end there. A vaudeville actress named Charlotte Bridgwood, known by her stage name Lotta Lawrence, picked up where Anderson left off and patented an electric wiper system that used rollers instead of blades and was powered by the car's engine.

However, neither Anderson nor Bridgwood received credit or any money for their contributions.

In some versions of windshield wipers' history, it was Bridgwood's daughter, Florence Lawrence, who invented electric windshield wipers and had her mother help her get the patent.

Windshield wipers finally found their place on automobiles in 1913 when Ford began adding them to the Model T, and Cadillac made wipers standard on their vehicles in 1922.

DID YOU KNOW?

Florence Lawrence is known as "the first movie star," and when she was able to afford her first automobile, she became engrossed in learning everything about it and how it worked. She was one of the first to create turn signals, with the use of a button to raise a flag in the back of the car, indicating which direction she would turn.

She also created a mechanism similar to a brake signal that she could deploy for a small Stop signal to pop up in the back of her car to warn motorists behind her when she was coming to a halt.

Similar to Anderson and Bridgwood, Lawrence's signal idea was copied too often and marketed by others, leading Lawrence to receive no compensation.

TL;DR VERSION

- Before windshield wipers, drivers either had to stick their heads out the window to see in inclement weather or stop the vehicle and remove the debris by hand.
- Many came up with similar ideas for windshield wipers, but the concept was lost on car manufacturers, while others worried wipers would create a distraction for drivers.
- Florence Lawrence is not only known as "the first movie star," but also created an early turn signal and braking signal mechanism. She may also have been behind the first electric windshield wiper idea.

PART EIGHT:
TECHNOLOGY

CALLER ID

Gather 'round, kids. This starts off a bit on the scary side, so grab your marshmallows and a buddy as I tell this spooky tale.

Back before everyone, it seems, carried their own phones in their pockets at all times, whole households would share just one phone. Yes, that meant everyone in the family had the same number.

For many, that phone was attached to the wall (believe it or not, kiddies). Before the subject of this chapter was introduced nationwide, when a phone rang, you would have no idea who was on the other end. It was a constant gamble. It could be a family member or friend for a chat (phew!), or it could be…a TELEMARKETER!

Nowadays, most people I know don't pick up calls from numbers that are unknown to them, but, back then, we had no way of seeing what phone number was on the other end of that ringing. For the mischievous, it was a wonderful time, when prank calling was considered a fun pastime for kids—viewers of the early seasons of *The Simpsons* probably remember Bart's many prank calls to Moe's Tavern.

As a kid back in this time, I was given butler-like duties when the phone rang. I had a script I followed. Even though it was rare that a phone call was ever for me until I got into middle school, I still raced for the phone before I knew that many calls would be like the junk mail of telecommunications.

"Hello, this is Emily, may I ask who's calling?"—I still shudder thinking about those words now. Luckily, the '90s smiled on my generation and Caller ID became commonplace in homes, although many were disappointed at the idea of no more prank phone calls, unless you knew the secret code to punch in before dialing (*67 was the code, by the way—not that I remember that off the top of my head, because I never made prank calls. No, not me…).

In a time with text messages, direct messages on social media platforms, and email, phone calls are becoming less and less the norm. Before all this, we had one saving grace: Caller ID.

The technology behind it was developed by Theodore George Paraskevakos beginning in 1968. When he developed a transmitter and receiver with the capability to recognize phone numbers of incoming calls, phone companies jumped at the technology. Originally, they wanted to capitalize on it by charging customers a per-use charge and giving users a voice announcement for each new call.

Dr. Shirley Ann Jackson, the first African-American to earn a PhD from the Massachusetts Institute of Technology, also contributed to Caller ID by creating technology used in, not only that, but also Call waiting and the portable fax.

However, John Harris had the idea to incorporate a screen into a telephone unit to display the caller's telephone number. He worked with Kazuo Hashimoto to build the prototype. Users would have to program the names associated with familiar numbers by hand, but it was still a step in the right direction. Originally, the idea for this was for residential areas, with business use as a secondary application.

Harris also developed the technology to change the ringtone based on who was calling, and even stop ringing when someone you're ignoring calls. He even added a feature to make the phone ring at a certain time as an alarm.

Soon, telephone companies got on board with the technology and, using phone book directories, Caller ID could display the name associated with that number, even if it wasn't pre-programmed into the telephone or Caller ID box. Caller ID began showing up in households as early as 1984.

When Call Waiting was introduced in 1995, it made Caller ID even more useful for determining if the call you were currently on was more or less important than the new incoming call.

Now that many people use cell phones as their main form of communication, house phones have fallen out of favor, meaning listed phone numbers have also declined.

Some advanced Caller ID systems can still display the name of the person associated with a cell phone number, although most cell-to-cell calls don't use this technology as often. We now program names and numbers into our cell phones, although it's much easier than it was when Harris originally introduced the technology. Back then, one would have to use the letters associated with the number keys to program in a name (think texting on your old Nokia in the early 2000s).

There are a few apps that can detect potential spam calls now. It's fascinating to think we've gone back to a time before Caller ID, and it's still common to text someone and receive the meme-able response of "New phone, who dis?"

DID YOU KNOW?

Caller ID is a brand name, although it's mostly treated as generic at this point.

TL;DR VERSION

- Before Caller ID, people just picked up the phone when it rang without even knowing who was on the other end!

- At first, phone companies wanted to use (and charge for) the technology on a per-call basis. Luckily, phones with displays and separate display units that could be connected to house phones were introduced to the market.
- Now that cell phones are more common than landline phones, phone numbers are not as frequently listed in directories, meaning Caller ID is not as effective as it once was.

KEVLAR

It's a common misconception that there's such a thing as a bulletproof vest. Just like there will never be a true all-in-one cleaner or the Number One cure for the common cold, most vests can be effective only against certain kinds of ammunition, but none will ever be effective against all.

However, that didn't stop inventors as far back as the 1500s from trying, and in the process, saving a multitudes of live. Metal was used in early armor to deflect bullets. Unfortunately, not only was the metal armor heavy and cumbersome, it also wasn't very effective, and many firearms could penetrate it. Nevertheless, this is when the term "bulletproof" emerged.

Japan and Korea began testing the effectiveness of silk in the 1800s. Though a soft material, it's also very tightly woven, and they both found that layering the material made it able to withstand being hit by the bullets of that era. A Polish monk living in Chicago heard about this technique and sought to improve upon it. Casimir Zeglen came to America at a time when public figures were falling victim to attacks by anarchists. In fact, the mayor of Chicago at the time, Carter Harrison Sr., was murdered in his home.

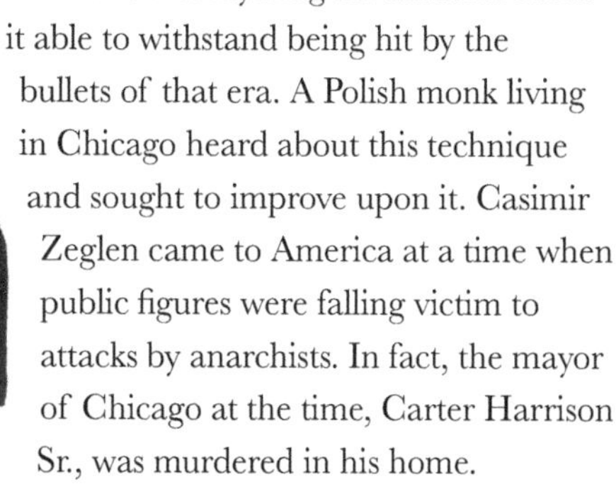

Zeglen wanted to devise a lightweight vest public figures could wear under their clothes on a day-to-day basis. He began experimenting with using a steel plate between layers of silk. In front of a live audience in New York City, he volunteered himself to test out his "bullet proof cloth" by being shot at a distance of only ten paces, to which he said he felt only a tap.

His invention was tested by the US military and, unfortunately, was found too hot and expensive, but Zeglen was determined to see his invention put to good use and offered the vest to President William McKinley. Before he could meet with the president, though, McKinley was fatally shot in Buffalo, New York. It turns out that, not only would Zeglen's vest have been able to stop the bullet used, but also, the vest would have protected the abdomen, where McKinley was shot.

Unfortunately, it's said Zeglen's vest wasn't able to stop "the shot heard around the world," which damaged his reputation and the reputation of the vest itself. Archduke Franz Ferdinand supposedly owned one of Zeglen's vests and, depending on the source you read, either was not wearing it when he was shot, or was hit in an area not covered by the vest he was wearing.

But the biggest innovation in the quest to find a material that could withstand the impact of a bullet came when Stephanie Kwolek took a job with DuPont in 1946, thinking it would be only temporary before she began medical school.

However, she found the work so fascinating, she stayed on with the company. When she was tasked with finding a material capable of replacing steel tire belting, she stumbled upon a synthetic plastic that was exponentially stronger than steel.

Kevlar was patented in 1966 and was first used in protective vests for police officers. The material has a couple hundred more applications, though, including coverings for underwater cables, building materials, gloves to protect hands from cuts, firefighter boots, hockey sticks and, as originally intended, car tires.

The next step in the ongoing search for the perfect bulletproof material is something called Spider Silk, made from genetically-engineered silkworms. It's still in the early stages of rigorous testing, but scientists are optimistic about its potential.

DID YOU KNOW?

If you think bulletproof vests are ugly, you're not alone. Colombian fashion designer Miguel Caballero became known as the "Armani of armor" for his bulletproof clothes fashioned to look like everyday outerwear. Famous clients include Steven Seagal and President Barack Obama.

TL;DR VERSION

- There's no such thing as a truly bulletproof vest.
- Early vests that were made to withstand a bullet were made of silk, followed by silk and steel.
- Kevlar was invented by Stephanie Kwolek at DuPont while she was working to create a material strong enough for tires.

PODCASTS

If you're a podcast listener (hopefully you at least know of one you like, if you're reading this book), you may have told others about shows you listen to, only to be met with a blank stare and the question, "What's a podcast?" It's not uncommon, since 40 percent of Americans still are not familiar with the term, which is a portmanteau of the words "iPod" and "broadcast."

The term "podcasting" is thought to come from a 2004 *Guardian* article, "Audible Revolution," by Ben Hammersley. The word was actually thrown into the article because Hammersley's editor told him the story was twenty words short. In his revision, he added: "But what to call it? Audio blogging? Podcasting? GuerillaMedia?"

The word "podcasting" stuck, and Hammersley is credited with coining the term. A few have rebelled against the name, like Leo Laporte from *This Week in Tech,* who calls himself a podcaster but believes the term "netcast" more accurately describes the medium, since not everyone listens on an iPod. However, considering Apple, Microsoft, Yahoo, and Google all support the name, it's here to stay.

Podcasting is possible because of the inventions of the RSS feed and the MP3 file. (Don't worry, I won't get too techy here. Stick with me.)

When you hear a podcaster say, "Subscribe to my podcast," what they're actually saying is, "Subscribe to my RSS feed, then, whenever a new MP3 file is uploaded, you will get instant access to it, and it will download automatically to whatever app or service you use to listen to podcasts."

RSS stands for Really Simple Syndication, by the way. And MP3 stands for Moving Picture Experts Group Layer-3 Audio, which is a coding format that uses lossy data compression, meaning it lowers the size of the file without sacrificing the quality distinguishable by the naked ear. (That's the extent of tech-geekiness for this, I promise.)

Some even say you cannot call an audio show a podcast unless it has an RSS feed.

By the way, RSS feeds are not just for podcasts—they're used in blogs and other websites that are updated, which you can subscribe to using an aggregator like Feedly so that any updates to the website you will see automatically, instead of having to refresh your favorite sites manually.

Software engineer Dave Winer is known as the man behind the RSS feed, while it was Adam Curry, a former MTV VJ, who is said to have popularized it for use with audio files.

Curry saw the potential of the internet as a media platform as far back as 1993, when he registered the domain MTV.com before the network did, resulting in a lawsuit the next year. But once he had the idea for an internet radio show, he was in contact with Winer about utilizing the RSS technology to make the internet "everyman's broadcast medium" without compromising quality, which was no easy feat then.

Christopher Lydon, who worked with Winer and now hosts the *Open Source* podcast, is known as the first voice to be released as a downloadable MP3 on an RSS feed in 2003, recording interviews for his blog. Winer is said to have come out with his show, *Morning Coffee Notes*, just a day before Adam Curry premiered *Daily Source Code* in August 2004. Lydon has said podcasting came about in response to the Iraq war, as a means to remedy the breakdown in the American and world conversation. Now it's used for people to listen to people talk about nearly every niche interest you can imagine (including quick bursts of the history of everyday objects, wink-wink).

Every few months, it seems, an article will come along that proclaims podcasting is the "next big thing" or the "Wild West" of the internet. But the technology has been around for quite a few years, and even at the annual conference for podcasters, Podcast Movement, the Academy of Podcasters induct hand-picked podcasters who have been in the space for at least ten years into the Podcasters Hall of Fame.

Podcast listening has also been growing year after year, but one of the great things is that, once people discover podcasts, especially ones they love, they want to tell everyone about them, kind of like, "Where has this been all my life?"

It can be said that, whenever a blogger, celebrity, entertainment network, or media outlet starts a podcast, they bring with them a portion of their already-existing audience, like *The Ricky Gervais Show* back in 2005, or Kevin Smith's *SModcast* in 2007, or *WTF* with Marc Maron in 2009. Even for independent podcasts, this is a great thing, since the more people find out about podcasts, the better it is for the industry.

When Apple introduced the Podcasts App for iPhone in 2012, even more people discovered these amazing little portable audio shows.

One of the biggest marks in the timeline of podcasts came in 2014 when Chicago Public Media's *This American Life* began hyping a spin-off show with a format of following a single story for a season of episodes, aptly called *Serial.*

Each year, podcast consumers seem to be looking for their next big fix, the one mainstream media will shine a spotlight on, but, for the most part, podcasters themselves see some of their best growth through grassroots and word of mouth from their listeners.

Podcasters and podcast listeners can even celebrate this audio awesomeness on September 30 every year, on International Podcast Day.

DID YOU KNOW?

You can listen to the podcast versions of many of the chapters in this book by searching your favorite podcast app for *The Story Behind* or by going to TheStoryBehindPodcast.com.

TL;DR VERSION

- The word "podcasting" can be traced back to 2004.
- Dave Winer and former MTV VJ Adam Curry are considered the creators of podcasting.
- International Podcast Day is celebrated by podcasters and podcast listeners every year on September 30.

VOICE RECOGNITION

Voice recognition began in 1952, when Bell Laboratories developed a machine that could recognize the spoken numbers zero through nine.

This was called "Audrey," an acronym of the words "AUtomatic Digit REcognizer" (plus a Y at the end so it would match the common name). And it was huge. Literally huge—the relay rack alone was six feet tall.

Audrey didn't talk back, but she responded with flashing lights, and she wouldn't respond to just anyone. In order to maintain the 97-percent accuracy of the machine, the speaker had to become acquainted with Audrey, and "she" was constantly fine-tuned and adjusted. All this for ten numbers to be recognized.

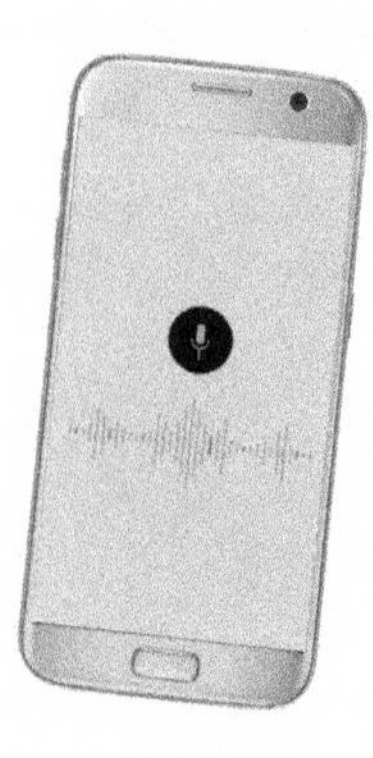

To take the analogy to teaching a baby to speak further, IBM introduced the Shoebox machine at the 1962 World's Fair, which understood a whopping sixteen English words. But it was the Department of Defense in the '70s that really advanced the technology of speech recognition. A department called the Defense Advanced

Research Projects Agency (DARPA) created "Harpy," voice recognition software that could understand more than one thousand words.

This was a huge first step to the voice recognition we're familiar with today. It was still a far cry from what science-fiction TV and movies, such as "Star Trek" and "Star Wars," were portraying then, but it was probably those movies that inspired the advancement to machines that could not only understand what we were saying, but also respond.

And now, because sometimes the '80s are scarier than anything we could have come up with in science fiction, let me introduce you to your new nightmare, Julie, the doll that could respond to your voice. The commercial was nightmare-worthy, at best, and downright horrifying at worst.

Yet, to a kid, it was quite possibly the most magical doll ever. This was right after the Cabbage Patch Kids craze, so, as you can imagine, doll companies were trying to capitalize on promoting different kinds of dolls. But the technology wasn't quite up to snuff, and many kids complained that Julie wasn't as responsive as she was portrayed to be in original commercials.

One thing to note about voice recognition is that the machines are not actually recognizing whole words—they are recognizing the smallest elements of spoken words, known as phonemes, and putting them together within the software to determine what word you are most likely to be saying.

Phonemes are the sound elements of speech, such as vowel sounds or plosives (harder consonant sounds).

The software may not always catch every phoneme, nor is it always able to format your speech and translate it to something the computer can understand—just think of the examples of yelling at Siri for screwing up a talk-to-text message or, as happens in our house, our Google Home recognizes when my husband says to turn off the living room lights, but not as often when I say it.

In 1990, a company called Dragon introduced the first speech-recognition program available to everyday consumers, called Dragon Dictate. For the low price of only $9,000, you could talk and have the program type what you were saying.

Even as the years progressed and the technology improved and became more affordable, seven years later, when Dragon NaturallySpeaking came out, the program could only recognize 100 words per minute, but the average rate of speaking is about 163 words per minute. So you would have to say a word, pause, say another word and pause, over and over again.

That was even after the initial set-up you needed to go through with the program that took about 45 minutes for it to learn your voice and speaking patterns. Even at almost $700, it was only 80 percent accurate.

DragonDictate is still around, by the way. In fact, getting back to science fiction, writer Peter David, known for writing for numerous comic book series, including "Star Trek" comic books, began using it in 2012 after his stroke.

One of the most progressive innovations in voice recognition came about with the introduction of MovieFone.

The words "Welcome to MovieFone" may bring back memories of finding out what movie to see for date night. Or you might remember them from memorable episodes of *Seinfeld* or *Family Guy*. Well, you can thank BellSouth for introducing the first voice portal, called VAL, short for Voice Activated Link, in 1996. This technology could be used over a telephone for "quick" (by the standards of the era) access to information about anything from movie times to restaurant suggestions to troubleshooting and customer service.

Being able to talk and get a spoken-word response was the piece of technology that really astounded us, though. But did you ever notice that many computer voices, such as navigation software, Siri, and the newer voice-activated assistants, have female voices?

This is intentional. Think back to the computers featured in the movies *2001: A Space Odyssey*, *WarGames*, and even the autopilot computer voice from *Wall-E*. They were all male and they were all on the side of evil in those portrayals. It's no wonder we're more comfortable having a female computer voice, when movies exacerbated our fears of computers with menacing male voices outsmarting humans.

However, one alternative to Google Home and Amazon Echo is voice-controlled software called Home Automated Living, otherwise known as HAL—you would think they would have considered the supercomputer, the HAL 9000, from *2001: A Space Odyssey* before picking that little nickname, though.

Even though now the voices of voice-activated assistants like Amazon Echo or Google Home are synthesized, there is supposedly a real person behind the original voice of Siri. Susan Bennett is a voice actor from Atlanta who took a job in 2005, recording different phrases for four hours a day for a month, for an unknown project from software company ScanSoFort When colleagues began playing with Siri in 2011, they recognized her voice, and she was just as surprised as they were. Apple will not confirm the voice of Siri, though, and it has since been updated to a different voice.

DID YOU KNOW?

By the way, if you do a Google search for the infamous *WarGames* line of "Shall we play a game?," the first result will be a game of tic-tac-toe you can play with Google.

TL;DR VERSION

- Voice recognition began in 1952 with a very large computer that recognized the numbers zero through nine and responded with blinking lights.

- Voice-recognition machines and software are not necessarily recognizing whole words, but parts of words called phonemes.
- Most voice-activated home automation tools use a female voice because it's considered less threatening.

PART NINE:

WEAPONS

GUNPOWDER

The origin of gunpowder starts, believe it or not, with the search for the elixir of life.

As early as 300 AD, Chinese scientists were experimenting with saltpeter, a chemical compound of potassium nitrate formed by decomposing animal manure—so, basically, they were playing with animal poop. By the eighth century, saltpeter and sulfur were combined with charcoal and the resulting concoction was used to treat skin diseases and kill insects. This was before anyone knew exactly how powerful the combination of those three ingredients could be.

Alchemists began using this combination of substances in their search for the elixir of life, but they found that it burned incredibly quickly and with tremendous intensity. A single spark could result in a fast and smoky fire. This dangerous combination was then used as part of the celebration of the emperor's birthday with the first known fireworks.

The Chinese knew that something as powerful as gunpowder could be used for much more. The Sung Dynasty began using it to defend themselves against the constantly invading Mongols. They placed gunpowder into a tube, followed by an arrow. When ignited from the other end, the gunpowder caught on fire and the gas produced would create enough pressure to propel the arrow out and across enemy lines.

Following that success, the Chinese invented more gunpowder-based weaponry, including the first cannons and grenades. By the thirteenth century, the science of gunpowder spread to Europe and the Islamic world. During the Middle Ages, cannons became common weapons used during the many wars, including the Hundred Years' War and the Siege of Constantinople.

The first handgun appeared in the fifteenth century. As it was, it was a shrunken-down cannon, which created a new class of soldier—the infantry.

With each new advance in gunpowder weaponry came the advancement of defenses against them, meaning weaponry had to stay one step ahead, which perpetuated the arms race we're still seeing today.

You may be familiar with the Guy Fawkes masks popularized by the movie *V for Vendetta* and associated with the hacktivist group Anonymous. Guy Fawkes is the reason for Gunpowder Night—otherwise known as Fireworks Night, Bonfire Night, or Guy Fawkes Night. The holiday is celebrated in the UK on November 5, hence the phrase "Remember, Remember the Fifth of November."

Back in 1605, King James and Parliament were persecuting Roman Catholics. Robert Catesby (whose father had been persecuted by Queen Elizabeth I), Guy Fawkes, and their co-conspirators made a plan to store gunpowder below the Parliament building. When Guy Fawkes was found lurking in a cellar under the building, he was taken into custody and thirty-six barrels of gunpowder were found.

Asked why he went to such extremes, Fawkes replied, "A desperate disease requires a dangerous remedy."

Fawkes was tortured until he confessed to conspiracy to blow up the English Protestant Government and replace it with a Catholic government. Fawkes was sentenced to be hanged, drawn, and quartered in London, but before he could be executed, he jumped from the ladder to the gallows, broke his neck and died.

King James—the same King James who was behind the King James Bible—burned an effigy of Guy Fawkes, and England passed restrictive laws on Catholics, including taking away their right to vote.

Parliament established November 5 as a day of public Thanksgiving the next year, but it's evolved into a celebration of Guy Fawkes and what came to be known as the Gunpowder Plot—albeit with many burning effigies of the Pope, Guy Fawkes, or contemporary political figures in bonfires. The only place in the UK that does not celebrate is Guy Fawkes' former school, St. Peter's in York, which refuses to burn effigies of him out of respect. Referring back to one of gunpowder's earliest uses, fireworks are always set off on that night.

DID YOU KNOW?

The King James version of the Bible was commissioned by the same King James Fawkes was protesting.

TL;DR VERSION

- Gunpowder was created while trying to find the elusive elixir of life.
- Once gunpowder was found to ignite quickly for use in fireworks, the application of it in weapons soon followed.
- The Gunpowder Plot was a resistance movement to Parliament's persecution of Roman Catholics in England in 1605.

THE REVOLVER

This chapter is based on an episode of The Story Behind *podcast that was part of a series based on the weapons from the game Clue.*

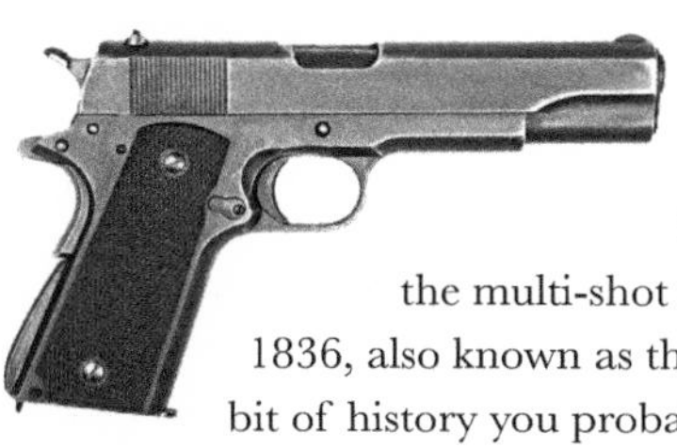

It's commonly known that Samuel Colt of Connecticut was granted a patent for the development of the multi-shot revolving firearm in 1836, also known as the revolver, but here's a bit of history you probably didn't learn about Colt and his invention: It was inspired by the wheel of a ship.

The use of a clutch to either lock the wheel or allow it to spin gave Colt the idea to apply the technique to a single-shot pistol to create a rotating cylinder of bullet chambers.

However, when he got this idea, he didn't have the money to create it, so he went on the road, performing a street show under the name Dr. Coult. And what exactly did he use to entertain crowds? Laughing gas.

His act was a success, and, with the money he saved, he was able to put assembly-line techniques from the Industrial Revolution to use in manufacturing his revolvers, which had become even more popular by the time the Civil War started.

Even without the war, the Colt revolvers were already well-known because of his promotion tactic of having famous artist George Catlin incorporate them into paintings. Think of this like the product placements you might see today in movies and television. He would also hire famous writers who traveled the world to write about them.

Probably the most well-known revolver is the Colt .45, also known as the Colt Single Action army handgun. Ironically, Samuel Colt never actually held one in his hands, since it wasn't released to the public until ten years after he died in 1862.

The Colt .45 was notably used in the shootout in Tombstone, Arizona, at the O.K. Corral in 1881, which included Wyatt Earp, Doc Holliday, and Billy Clanton. The gun became known as the Peacemaker or the Equalizer. Its legacy became widespread, and not only in dueling—the West itself was settled with the help of the Colt .45.

The gun was a favorite among heroes like Theodore Roosevelt's Rough Riders and George S. Patton; outlaws such as Billy the Kid, Jesse James, and Butch Cassidy; and even entertainers like Annie Oakley. With the popularity of early cowboy movies, the gun's popularity also grew, but when the West became less wild and new double-action revolvers and semi-automatic pistols came along, the popularity of the Colt .45 waned.

Manufacturing even slowed to a halt during World War II, but, with the resurgence of the Westerns of the 1950s, the Colt .45 made its way into the eye of the mainstream again. John Wayne even sported an ivory-gripped pair as his signature pistols. The popularity of television Westerns like *Bonanza* and *The Lone Ranger* helped boost Colt .45 manufacturing by 1956.

One more thing: The malt liquor called Colt .45 wasn't named after the Peacemaker, but rather in honor of 1963 Baltimore Colts running back Jerry Hill, whose jersey number was 45.

DID YOU KNOW?

The Beatles' album *Revolver* wasn't named for the weapon, but for the way an album spins on a record player.

TL;DR VERSION

- Samuel Colt was inspired by a ship's wheel to create the revolver.
- Colt was able to raise the money for manufacturing it through a touring act in which his main entertainment was the use of laughing gas.
- The Colt .45's popularity waned until Westerns and, in particular, John Wayne's signature pistols, became popular.

ACKNOWLEDGEMENTS

I would like to thank so many individuals for their support and encouragement, especially listeners and Patreon supporters of *The Story Behind* podcast who keep me podcasting week after week. I am grateful to everyone who was a part of the process of writing this book, especially those who saw me on social media when I should have been writing and cheered me on, while also gently kicking me off (you guys helped me make my deadline).

Thank you to the staff of Mango Publishing, especially Brenda Knight, who helped fulfill my childhood dream of becoming a writer; my podcasting mentors Jim Collison, Elsie Escobar, Jessica Kupferman, Dave Jackson, Stephen Jondrew, Daniel J. Lewis, Ray Ortega, Stargate Pioneer, Steve Stewart, and Rob Walch, who give me the motivation and know-how to get behind a microphone and explore my creativity and geeky side; my Mastermind group who kept me accountable for getting this book done, Jason Bryant, John Bukenas, Darrell Darnell, Mark Des Cotes, Wayne Henderson, Dave Jackson (listed yet again, but always worth thanking multiple times), and Laura McClellan; my Philly Fort Buddies, who offered encouraging glasses of wine, coffee breaks, and mental health check-ins along the way,

Josh Hallmark, Hannah Ostic, and Steven Pappas; Landon and Ash Porter, and the Gorillas who kept my Impostor Syndrome from constantly trying to sabotage me; but none of this would have been possible without the first and dearest friend I made in podcasting, Epic Film Guy Nicholas Haskins, who got me back behind the mic to start *The Story Behind*.

A note of thanks for friends and family who have always encouraged me as a writer and who put up with the random trivia I would spout out during awkward dinner party conversations before I had an outlet for that sort of stuff.

I would also like to offer my everlasting gratitude to my husband, Mark Prokop. Not only was he the one who suggested the idea of starting a podcast in the first place, but he also let his geeky side (and credit card) take over when it came to equipping me with everything I needed. He also offered his IT services whenever something wasn't working and his copy-editing skills in the final hours before this book was due. He is also an amazing father, and I'll never be able to match the number of diapers he has changed while he has gracefully allowed me to find my passion in podcasting and in writing this book.

TL;DR VERSION

- Listeners and supporters of *The Story Behind* are awesome!

- Those who believed in me and cheered me on are some of the best people in the world.
- My husband is the Ben to my Leslie.

THE PODCAST

Many of the chapters in this book are based on episodes of the podcast *The Story Behind*, which is available on Apple Podcasts, Google Podcasts, Spotify, PocketCasts, I Heart Radio, and even YouTube. Visit TheStoryBehindPodcast.com to find out how to subscribe and also listen to the audio versions.

The following episodes from the book are available:

Band-Aids.........TheStoryBehindPodcast.com/BandAids

Bubble Gum.........TheStoryBehindPodcast.com/BubbleGum

Comic Sans.........TheStoryBehindPodcast.com/ComicSans

Dr Pepper.........TheStoryBehindPodcast.com/DrPepper

Gunpowder.........TheStoryBehindPodcast.com/Gunpowder

Hypnosis.........TheStoryBehindPodcast.com/Hypnosis

The Lead Pipe.........TheStoryBehindPodcast.com/LeadPipe

The Lollipop.........TheStoryBehindPodcast.com/Lollipop

Lullabies.........TheStoryBehindPodcast.com/Lullabies

Mad Hatters.........TheStoryBehindPodcast.com/MadHatters

Musicals.........TheStoryBehindPodcast.com/Musicals

Mustaches.........TheStoryBehindPodcast.com/Mustaches

The Paper Clip.........TheStoryBehindPodcast.com/PaperClip

Peanut Butter.........TheStoryBehindPodcast.com/PeanutButter

Ping-Pong.........TheStoryBehindPodcast.com/PingPong

The Pink Slip.........TheStoryBehindPodcast.com/PinkSlip

Pockets.........TheStoryBehindPodcast.com/Pockets

Podcasts.........TheStoryBehindPodcast.com/Podcasts

Pretzels.........TheStoryBehindPodcast.com/Pretzels

The Revolver.........TheStoryBehindPodcast.com/Revolver

Sliced Bread.........TheStoryBehindPodcast.com/SlicedBread

The Smiley Face.........TheStoryBehindPodcast.com/SmileyFace

Tattoos.........TheStoryBehindPodcast.com/Tattoos

The Theremin.........TheStoryBehindPodcast.com/Theremin

Tissues.........TheStoryBehindPodcast.com/Tissues

The Toothbrush.........TheStoryBehindPodcast.com/Toothbrush

The Traffic Light.........TheStoryBehindPodcast.com/TrafficLight

The Treadmill.........TheStoryBehindPodcast.com/Treadmill

The T-Shirt.........TheStoryBehindPodcast.com/Tshirt

Voice Recognition.........TheStoryBehindPodcast.com/Voice Recognition

EMILY PROKOP

Emily Prokop graduated from Southern Connecticut State University with a degree in journalism and has worked in newspapers, magazines, and even as a puzzle editor, but always loved picking up random trivia and sharing it with her not-quite-as-enthused friends and coworkers. Luckily, she was able to combine her passions of trivia and podcasting when she created *The Story Behind* in 2016. She lives in Connecticut with her husband, two kids, and two cats.